Advanced Mathematics and Computational Applications in Control Systems Engineering

Advanced Mathematics and Computational Applications in Control Systems Engineering

Editors

Francisco-Ronay López-Estrada
Guillermo Valencia-Palomo

MDPI • Basel • Beijing • Wuhan • Barcelona • Belgrade • Manchester • Tokyo • Cluj • Tianjin

Editors

Francisco-Ronay López-Estrada
TURIX-Dynamics Diagnosis and Control Group
Tecnológico Nacional de México - IT Tuxtla Gutiérrez
Tuxtla Gutiérrez
Mexico

Guillermo Valencia-Palomo
Graduate Studies and Research Division
Tecnológico Nacional de México - IT Hermosillo
Hermosillo
Mexico

Editorial Office
MDPI
St. Alban-Anlage 66
4052 Basel, Switzerland

This is a reprint of articles from the Special Issue published online in the open access journal *Mathematical and Computational Applications* (ISSN 2297-8747) (available at: www.mdpi.com/journal/mca/special_issues/ctrl_syst_eng).

For citation purposes, cite each article independently as indicated on the article page online and as indicated below:

LastName, A.A.; LastName, B.B.; LastName, C.C. Article Title. *Journal Name* **Year**, *Volume Number*, Page Range.

ISBN 978-3-0365-1452-9 (Hbk)
ISBN 978-3-0365-1451-2 (PDF)

© 2021 by the authors. Articles in this book are Open Access and distributed under the Creative Commons Attribution (CC BY) license, which allows users to download, copy and build upon published articles, as long as the author and publisher are properly credited, which ensures maximum dissemination and a wider impact of our publications.
The book as a whole is distributed by MDPI under the terms and conditions of the Creative Commons license CC BY-NC-ND.

Contents

About the Editors

Francisco-Ronay López-Estrada

Francisco-Ronay López-Estrada received his PhD degree in automatic control from the University of Lorraine, France, in 2014. He has been with the National Technological Institute of Mexico, IT Tuxtla Gutiérrez, as a lecturer since 2008. He is part of the editorial board of the journals *Mathematical and Computational Applications* and the *International Journal of Applied Mathematics and Computer Science*. He is the author/co-author of more than 40 research papers published in ISI journals and several international conferences. He has led several funded research projects, and the funding comprised a mixture of sole investigator funding and collaborative grants. Furthermore, he collaborates with European universities such as the University of Lorraine, University of Paris 2, University of Stavanger, and Polytechnique University of Catalonia, among others. His research interests are fault diagnosis and fault-tolerant control based on convex LPV and Takagi–Sugeno models and their applications.

Guillermo Valencia-Palomo

Guillermo Valencia-Palomo received his Eng. degree in electronics from the Tecnológico Nacional de México-IT Mérida, in 2003, and his MSc and PhD degrees in automatic control from Cenidet in 2006 and The University of Sheffield, UK, in 2010, respectively. Since 2010, he has been with Tecnológico Nacional de México-IT Hermosillo. He is author/co-author of more than 40 papers published in ISI journals, several international conferences, and one patent. He has led a number of funded research projects, and the funding for these projects comprised a mixture of sole investigator funding, collaborative grants, and funding from industry. He is associate editor of *IEEE Access*, the *International Journal of Aerospace Engineering*, and *IEEE Latin America Transactions* and a member of the editorial board of *Mathematical and Computational Applications*. His research interests include predictive control, LPV systems, fault detection, fault-tolerant control systems, and their applications.

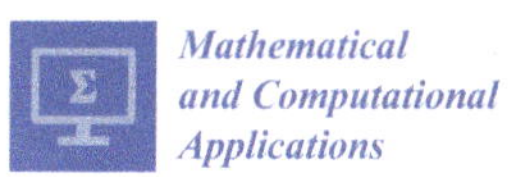
Mathematical and Computational Applications

MDPI

Editorial

Advanced Mathematics and Computational Applications in Control Systems Engineering

Francisco-Ronay López-Estrada [1] and Guillermo Valencia-Palomo [2,*]

1 Tecnológico Nacional de México, IT Tuxtla Gutiérrez, TURIX-Dynamics Diagnosis and Control Group, Carr. Panam. km 1080 S/N, Tuxtla Gutierrez 29050, Mexico; frlopez@iitg.edu.mx
2 Tecnológico Nacional de México, IT Hermosillo. Av. Tecnológico y Periférico Poniente, S/N, Hermosillo 83170, Mexico
* Correspondence: gvalencia@hermosillo.tecnm.mx; Tel.: +52-662-260-6500 (ext. 230)

Citation: López-Estrada, F.R.; Valencia-Palomo, G. Advanced Mathematics and Computational Applications in Control-Systems Engineering. *Math. Comput. Appl.* **2021**, *26*, 20. https://doi.org/10.3390/mca26010020

Received: 2 March 2021
Accepted: 2 March 2021
Published: 3 March 2021

Publisher's Note: MDPI stays neutral with regard to jurisdictional claims in published maps and institutional affiliations.

Copyright: © 2021 by the authors. Licensee MDPI, Basel, Switzerland. This article is an open access article distributed under the terms and conditions of the Creative Commons Attribution (CC BY) license (https://creativecommons.org/licenses/by/4.0/).

Control-systems engineering is a multidisciplinary subject that applies automatic-control theory to design systems with desired behaviors in control environments. Automatic-control theory has played a vital role in the advancement of engineering and science. It has become an essential and integral part of modern industrial and manufacturing processes. Today, control-precision requirements are higher, and real systems are more complex, including higher-order, discrete, hybrid, time-delayed linear and nonlinear systems, and systems without a mathematical model and uncertainties.

In control engineering, parallel to all other engineering disciplines, the impact of advanced mathematical and computational methods is rapidly increasing. Advanced mathematical methods are needed because real-world control systems need to comply with several conditions related to product-quality and -safety constraints that have to be considered in problem formulation. Conversely, the increment in mathematical complexity impacts computational aspects related to numerical simulation and practical implementation of algorithms. A balance must also be maintained between implementation costs and the performance of the control system.

This special issue aims to present recent advances in developing and applying advanced mathematics and computational applications in control-systems engineering. It comprises nine high-quality papers, summarized below.

Alhato and Bouallègue [1] present an advanced metaheuristic optimization algorithm—thermal-exchange optimization—in order to optimize the gains of proportional–integral (PI) controllers applied to outer loops in the classical vector control scheme of doubly fed induction-generator-based wind-turbine systems. The authors considered the PI controllers' gain tuning as an optimization problem under nonlinear and nonsmooth operational constraints to improve the precision of the proposed control technique in reference tracking.

Kas and Das [2] provide a new theoretical approach of controlling resistance spot welding processes with the aim of avoiding inconsistent weld quality and inadequate nugget size. The proposal is based on a dynamical analytical model and an adaptive tracking controller that continuously adjusts welding voltage given the estimation of unknown process parameters.

Lara-Ortiz et al. [3] developed a hybrid active disturbance rejection controller for the regulation of the gait cycle of a mobile worm-bioinspired robot. The controller considers the maximal and minimal angles of each of the six links that conform the worm structure. An extended state observer estimates the unknown dynamics to actively compensate for it in a closed loop. Both numerical and experimental results validated the strategy.

Rodríguez-Mata et al. [4] present an approach based on Streeter–Phelps contaminant-distribution models and state-estimation techniques to monitor water-quality conditions. They designed a novel fractional-order high-gain observer to estimate dissolved- and biochemical-oxygen demand, which are variables that are difficult to measure with physical sensors. Experimental results of its application in a river are presented.

López-Estrada et al. [5] studied the case of a mechanical crane transporting a load of which the control objective was to attenuate swing load while tracking a reference. They propose a convex quasilinear parameter-varying approach with a $\mathcal{H}_\infty$ criterion that guarantees robustness against measurement noise and partial faults.

Pulido-Luna et al. [6] discuss a method to synchronize unidirectionally coupled heterogeneous chaotic systems, and its implementation in reconfigurable hardware. The controller uses Lyapunov theory with state feedback for this purpose. Lastly, it was implemented in a field-programmable gate array (FPGA) to test its performance.

Heras-Cervantes et al. [7] dealt with a DC–DC buck–boost power converter to control the heat of a resistance located in the boiler of a distillation column. Three different models of the converter (switching, nonlinear, and fuzzy Takagi–Sugeno) and two different fuzzy observers (with and without sliding modes) to estimate the inductor current and the capacitor voltage were compared in order to determine the best performance option.

Santos-Ruiz et al. [8] propose a method to simultaneously estimate the head loss and roughness of a serpentine pipe through the use of nonlinear optimization. The authors optimized the error of the Colebrook–White equation for an operating interval using the pressure and flow measurements at the pipeline ends.

Ramírez-Cárdenas and Trujillo-Romero [9] focus on sensorless speed tracking of a brushless direct-current motor using a neural network. The neural network had two layers; it was trained using the backpropagation method, and the velocity values and control signal from a proportional–integral–derivative control. The neural network could hold constant and variable load pairs, as is numerically demonstrated.

Lastly, we would like to acknowledge the enthusiastic authors, reviewers, and editorial staff who made this Special Issue possible.

Conflicts of Interest: The authors declare no conflict of interest.

References

1. Alhato, M.M.; Bouallègue, S. Direct power control optimization for doubly fed induction generator based wind turbine systems. *Math. Comput. Appl.* **2019**, *24*, 77. [CrossRef]
2. Kas, Z.; Das, M. Adaptive Control of Resistance Spot Welding Based on a Dynamic Resistance Model. *Math. Comput. Appl.* **2019**, *24*, 86. [CrossRef]
3. Lara-Ortiz, V.; Salgado, I.; Cruz-Ortiz, D.; Guarneros, A.; Magos-Sanchez, M.; Chairez, I. Hybrid State Constraint Adaptive Disturbance Rejection Controller for a Mobile Worm Bio-Inspired Robot. *Math. Comput. Appl.* **2020**, *25*, 13. [CrossRef]
4. Rodriguez-Mata, A.E.; Bustos-Terrones, Y.; Gonzalez-Huitrón, V.; Lopéz-Peréz, P.A.; Hernández-González, O.; Amabilis-Sosa, L.E. A Fractional High-Gain Nonlinear Observer Design—Application for Rivers Environmental Monitoring Model. *Math. Comput. Appl.* **2020**, *25*, 44. [CrossRef]
5. López-Estrada, F.R.; Santos-Estudillo, O.; Valencia-Palomo, G.; Gómez-Peñate, S.; Hernandez-Gutiérrez, C. Robust qLPV Tracking Fault-Tolerant Control of a 3 DOF Mechanical Crane. *Math. Comput. Appl.* **2020**, *25*, 48. [CrossRef]
6. Pulido-Luna, J.R.; López-Rentería, J.A.; Cazarez-Castro, N.R. Design of a Nonhomogeneous Nonlinear Synchronizer and Its Implementation in Reconfigurable Hardware. *Math. Comput. Appl.* **2020**, *25*, 51. [CrossRef]
7. Heras-Cervantes, M.; Téllez-Anguiano, A.d.C.; Anzurez-Marín, J.; Espinosa-Juárez, E. Analysis and Comparison of Fuzzy Models and Observers for DC-DC Converters Applied to a Distillation Column Heating Actuator. *Math. Comput. Appl.* **2020**, *25*, 55. [CrossRef]
8. Santos-Ruiz, I.; López-Estrada, F.R.; Puig, V.; Valencia-Palomo, G. Simultaneous optimal estimation of roughness and minor loss coefficients in a pipeline. *Math. Comput. Appl.* **2020**, *25*, 56. [CrossRef]
9. Ramírez-Cárdenas, O.D.; Trujillo-Romero, F. Sensorless speed tracking of a brushless DC motor using a neural network. *Math. Comput. Appl.* **2020**, *25*, 57. [CrossRef]

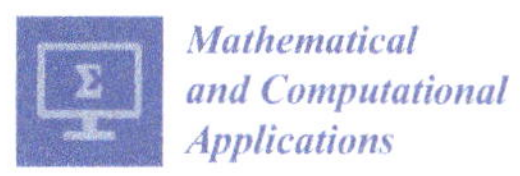

Article

Direct Power Control Optimization for Doubly Fed Induction Generator Based Wind Turbine Systems

Mohammed Mazen Alhato [1,*] and Soufiene Bouallègue [1,2,*]

[1] Research Laboratory in Automatic Control (LARA), National Engineering School of Tunis (ENIT), University of Tunis, El Manar, BP 37, Le Belvédère, 1002 Tunis, Tunisia

[2] High Institute of Industrial Systems of Gabès, 6011 Gabès, Tunisia

* Correspondence: mohammedhato.89@gmail.com (M.M.A.); soufiene.bouallegue@issig.rnu.tn (S.B.)

Received: 16 July 2019; Accepted: 22 August 2019; Published: 26 August 2019

Abstract: This study presents an intelligent metaheuristics-based design procedure for the Proportional-Integral (PI) controllers tuning in the direct power control scheme for 1.5 MW Doubly Fed Induction Generator (DFIG) based Wind Turbine (WT) systems. The PI controllers' gains tuning is formulated as a constrained optimization problem under nonlinear and non-smooth operational constraints. Such a formulated tuning problem is efficiently solved by means of the proposed Thermal Exchange Optimization (TEO) algorithm. To evaluate the effectiveness of the introduced TEO metaheuristic, an empirical comparison study with the homologous particle swarm optimization, genetic algorithm, harmony search algorithm, water cycle algorithm, and grasshopper optimization algorithm is achieved. The proposed TEO algorithm is ensured to perform several desired operational characteristics of DFIG for the active/reactive power and DC-link voltage simultaneously. This is performed by solving a multi-objective function optimization problem through a weighted-sum approach. The proposed control strategy is investigated in MATLAB/environment and the results proved the capabilities of the proposed control system in tracking and control under different scenarios. Moreover, a statistical analysis using non-parametric Friedman and Bonferroni–Dunn's tests demonstrates that the TEO algorithm gives very competitive results in solving global optimization problems in comparison to the other reported metaheuristic algorithms.

Keywords: doubly fed induction generator; PI tuning; LCL-filter; passive damping; advanced metaheuristics; Bonferroni–Dunn and Friedman's tests

---

1. Introduction

Recently, the increasing consumption of electrical energy, depletion of fossil fuels and the environmental problems related to using the non-renewable sources have promoted a growing interest in renewable energies [1,2]. Wind power as a renewable source represents an important and a promising solution for electrical demand and has recently gained more attention for its significant merits of cleanness and resource abundances. Various wind energy configurations are produced from the intensive studies and researches that are carried out in wind systems. One of the most common configurations is the grid-connected Doubly Fed Induction Generators (DFIGs) equipped with variable speed Wind Turbines (WTs). This configuration was widely installed in wind industry due to its significant merits such as independent control of active and reactive powers, low converters' costs and mechanical stress reduction [2].

A DFIG is a wound rotor generator in which the stator windings are directly connected to the grid. The rotor windings are connected to the grid through a back-to-back power converter that is composed from the Rotor Side Converter (RSC) and Grid Side Converter (GSC) components. This configuration allows the converter to handle the fraction of 20% to 30% of the total power. Therefore, the losses in the converter are lower compared to a system where the converter has to handle the full power leading to an improvement in the total efficiency [2,3].

However, the switching of Insulated-Gate Bipolar Transistors (IGBTs) at both the RSC and GSC components is usually performed by Sinusoidal Pulse Width Modulation (SPWM) strategy, which generates the high frequency content in the utility current. A simple L-filter or an LCL-filter is adopted to reduce the Total Harmonic Distortion (THD) of the utility current [4]. To meet the grid code requirements, the LCL-filter is predominant in reducing the utility current harmonics. Indeed, it can lead to a better attenuation of harmonics using small values of inductances. However, the resonance phenomena of the LCL-filter must be damped properly in order to prevent the possible instability of these systems [4]. One way for solving this problem is employing a passive damping circuit. The passive damping is built by inserting a resistor branch in series or in parallel with the inductors or capacitors of the LCL-filter [4].

The conventional control scheme of the grid connected DFIG wind turbine system is built based on a vector control method [5–8]. Here, the Stator Flux Orientation (SFO) scheme is selected for DFIG power regulation at the RSC. The SFO-based vector control strategy enables a decoupled regulation of the active and reactive powers that is flowing between the DFIG and the grid. The active and reactive powers' control strategy is performed by regulating the converter currents using Proportional-Integral (PI) controllers [5–8]. On the other hand, the Voltage Oriented Control (VOC) is adopted at the GSC, which the DC-link voltage is maintained to constant value [9].

Zhou and Blaabjerg [5] proposed a frequency-domain based approach to attain the PI controllers' gains for the inner and outer loops at both RSC and GSC circuits. The gains of the PI controllers that verify the design condition are selected for control loop. However, Hamane et al. [6] performed the direct power control of the DFIG using both PI and sliding mode controllers. The synthesis of the PI controller is based on an algebraic pole compensation method. Moreover, Hamane et al. [7] developed a comparative study of PI, RST, sliding mode and fuzzy supervisory controllers, and the synthesis of PI and polynomial RST controllers are tuned based on the pole compensation and pole assignment methods, respectively. In addition, the indirect vector control strategy is adopted, in which the matrix converter is used at the wound rotor of the DFIG instead of the conventional back-to-back converter in [8]. The active/reactive powers and current components are regulated by using the PI controllers that are designed using the pole placement technique. Although these methods have presented good performances, the main drawback for this type of control is that the performance of the DFIG system highly depends on a proper tuning of the PI controller gains. However, since the parameters of PI controllers depend on the precise mathematical representation, the control schemes become prone to error. In addition, the execution of tuning process can be time-consuming, and the optimal gains may not be obtained. Hence, proposing a systematic approach to find the best parameters setting of PI controllers is an interesting task and the metaheuristics-based hard optimization theory may be considered a promising solution.

In recent years, an enormous variety of metaheuristics optimization algorithms has been applied to solve complex and hard problems in science and engineering fields. Bekakra and Attous [10] presented the Particle Swarm Optimization (PSO) algorithm to obtain the optimal gains of PI controllers for the indirect control of the active and reactive powers' loops of a DFIG. Integral Absolute Error (IAE), Integral Time-weighted Absolute Error (ITAE) and Integral Square Error (ISE) performance criteria were selected as objective functions. Vieira et al. [11] applied the Genetic Algorithm (GA) to obtain the gains of the PI controllers at the RSC, where the active and reactive powers loops were directly controlled. In addition, an indirect power control strategy was achieved using GA-tuned integral sliding mode controllers, where the GA was used to tune the parameters of the direct and quadrature rotor current controllers [12]. Moreover, Assareh et al. [13] proposed a hybrid GA along with a gravitational search algorithm to attain the optimal gains of the PI controller for the torque regulation of a DFIG-based WT system. In all of these studies, the optimization algorithms are applied to obtain only the PI controllers' gains of the active and reactive powers at the RSC without discussion about the DC-link voltage regulation loop at the GSC and their effect on the overall control system.

Therefore, this work tends to apply a unified Thermal Exchange Optimization (TEO) metaheuristic algorithm to optimize the gains of PI controllers for the outer-loops in the classical vector control scheme of a DFIG-based wind energy system. In particular, this paper deals with the PI controllers tuning of the active and reactive power control loops in the RSC. It also treats the PI controller optimization-based design for the DC-link voltage loop in the GSC component.While the inner current loops at both RSC and GSC are tuned according to classical methods. The global TEO metaheuristic that has been proposed by Kaveh and Dadras in 2017 [14] is mainly adopted to tune the controllers' parameters of the modeled DFIG-based Wind Energy Converter System (WECS). Moreover, other well-known methods such as PSO, GA, Harmony Search Algorithm (HSA), Water Cycle Algorithm (WCA) and Grasshopper Optimization Algorithm (GOA) are used for an empirical comparison study. Moreover, a statistical analysis is performed to check the significance of each algorithm by using nonparametric tests such as Friedman's rank and Bonferroni–Dunn's test. The main contribution of this work is to provide a systematic and less complex procedure based on an advanced TEO algorithm in order to design and tune all outer-loops PI controllers for the well-known vector control scheme of a DFIG-based wind energy system. The classical trials-errors based methods of PI controller tuning are no longer used and the design time is further reduced. The drawbacks of the classical tuning methods are significantly reduced.

The remainder of this paper is organized as follows. The mathematical model of DFIG based WT is presented in Section 2. The description of the vector control scheme for the DFIG system at both RSC and GSC is analyzed in Section 3. In addition, Section 4 is devoted to the formulation of the outer-loops PI controllers' tuning problem given as a constrained optimization problem to be solved by the proposed TEO algorithm. In Section 5, such a TEO algorithm is described and its pseudo-code for the software implementation is given. Section 6 presents the implementation and the validation of the proposed TEO-tuned PI controller approach. Concluding remarks are given in Section 7.

2. Modelling of the DFIG Based Wind Energy Converter

The configuration of the DFIG-based WECS is depicted in Figure 1. A WT is joined to the DFIG by means of a gearbox. The DFIG is an induction machine, which the stator windings are directly connected to the grid, while the rotor windings are connected to the grid thanks to a back-to-back converter.

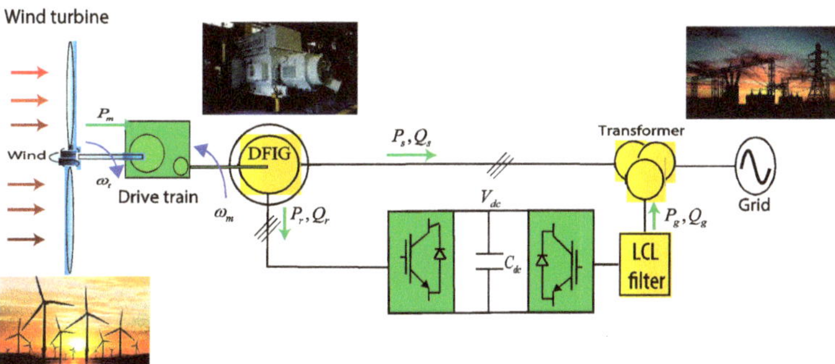

Figure 1. Configuration of the Doubly Fed Induction Generator (DFIG)-based Wind Energy Converter System (WECS).

2.1. Modelling of the Wind Turbine

The rotor blade of the WT is responsible for catching the wind power and converting it into kinetic energy. The captured mechanical power P_m is given as follows [2]:

$$P_m = \frac{1}{2}\rho C_p(\lambda, \beta)\pi R^2 V_w^3 \tag{1}$$

where ρ is the air density, C_p is the power coefficient and it depends on both the tip-speed ratio λ and the blade pitch angle β, R is the turbine radius and V_w is the wind speed.

The tip speed ratio is defined as the ratio of the blade tip speed to the wind speed. It is given by:

$$\lambda = \frac{\Omega_t R}{V_w} \tag{2}$$

where Ω_t is the angular speed of the WT.

2.2. Modelling of the DFIG

The dynamical model of the studied DFIG can be expressed by the following electrical equations [6–8]:

$$\begin{cases} V_{ds} = R_s i_{ds} + \frac{d\varphi_{ds}}{dt} - \omega_s \varphi_{qs} \\ V_{qs} = R_s i_{qs} + \frac{d\varphi_{qs}}{dt} + \omega_s \varphi_{ds} \\ V_{dr} = R_r i_{dr} + \frac{d\varphi_{dr}}{dt} - (\omega_s - \omega_m)\varphi_{qr} \\ V_{qr} = R_r i_{qr} + \frac{d\varphi_{qr}}{dt} + (\omega_s - \omega_m)\varphi_{dr} \end{cases} \tag{3}$$

where V_s and i_s are the stator voltage and current, V_r and i_r are the rotor voltage and current, φ_s and φ_r are the stator and rotor flux linkages, R_s and R_r are the stator and rotor resistances, ω_s and ω_m are the stator and rotor angular frequencies, respectively. The subscripts d and q denote the direct and quadrature axis components, respectively. The stator and rotor flux linkages are defined as follows:

$$\begin{cases} \varphi_{ds} = L_s i_{ds} + L_m i_{dr} \text{ and } \varphi_{qs} = L_s i_{qs} + L_m i_{qr} \\ \varphi_{dr} = L_r i_{dr} + L_m i_{ds} \text{ and } \varphi_{qr} = L_r i_{qr} + L_m i_{qs} \end{cases} \tag{4}$$

where L_s, L_r and L_m are the stator, rotor and magnetizing inductances, respectively.

2.3. Modelling of the GSC and the DC-Link

The GSC is connected to the grid through an LCL-filter. However, for better understanding the control of GSC, it is necessary to describe the model of the grid side system and DC-link parts. The mathematical formulation of the GSC in the dq synchronous frame is defined by Equation (5). In this equation, L_T represents the sum of the converter L_i and grid side L_g inductances. In fact, the LCL-filter can be approximated to an L-filter equal to the sum of the LCL-filter inductors [4]:

$$\begin{cases} L_T \frac{di_{gd}}{dt} = -R_T i_{gd} + \omega_g L_T i_{gq} + e_{gd} - V_{df} \\ L_T \frac{di_{gq}}{dt} = -R_T i_{gq} - \omega_g L_T i_{gd} + e_{gq} - V_{qf} \end{cases} \tag{5}$$

where V_{df} and V_{qf} are the components of the converter side voltage, i_{gd} and i_{gq} are the components of the grid currents, e_{gd} and e_{gq} are the components of the grid voltage, and R_T is the sum of the converter and grid side resistors.

The grid active and reactive powers are expressed as [9]:

$$\begin{cases} P_g = \frac{3}{2}\left(e_{dg} i_{dg} + e_{qg} i_{qg}\right) \\ Q_g = \frac{3}{2}\left(e_{qg} i_{dg} - e_{dg} i_{qg}\right) \end{cases} \tag{6}$$

3. Vector Control of the DFIG-Based Wind Energy Converter

The classical direct power control of the DFIG system is divided into the RSC and GSC control loops, where the control structure for the RSC and GSC components consist of two cascaded control layers. The inner PI layers are adopted to regulate the components of rotor and grid currents. On the other hand, the outer PI ones in the RSC are implemented to control the active and reactive powers while the outer PI loop in the GSC is adopted to maintain the DC-link voltage to its reference value. However, in this work, the inner PI loops in the RSC and GSC circuits are designed based on a pole assignment technique [8,15], while the outer PI ones are tuned based on the proposed metaheuristics algorithms-based approach.

3.1. Control of the RSC

The RSC circuit is responsible to independently regulate the active and reactive powers in order to extract the maximum available power. The vector control by the SFO scheme is employed for the power regulation of the DFIG system.

Considering that the electrical network is stable, the stator flux φ_s is constant. In addition, for the medium and high power DFIG, the value of the stator resistance is very small and therefore it can be neglected [6,7]. Hence, based on these assumptions, the stator voltages and fluxes in Equation (3) and Equation (4) can be rewritten respectively as follows:

$$\begin{cases} V_{ds} = 0 \\ V_{qs} = V_s = \omega_s \varphi_{ds} \end{cases} \tag{7}$$

$$\begin{cases} \varphi_{ds} = \varphi_{ds} = L_s i_{ds} + L_m i_{dr} \\ \varphi_{qs} = 0 = L_r i_{qs} + L_m i_{qr} \end{cases} \tag{8}$$

The active and reactive stator powers and rotor voltages are given, respectively, by:

$$\begin{cases} P_s = \frac{3}{2}\left(V_{ds} i_{ds} + V_{qs} i_{qs}\right) = -\frac{3}{2}\frac{L_m}{L_s} V_s i_{qr} \\ Q_s = \frac{3}{2}\left(V_{qs} i_{ds} + V_{ds} i_{qs}\right) = \frac{3}{2}\left(\frac{V_s^2}{\omega_s L_s} - \frac{L_m}{L_s} V_s i_{dr}\right) \end{cases} \tag{9}$$

$$\begin{cases} V_{dr} = R_r i_{dr} + \sigma L_r \frac{di_{dr}}{dt} - \overbrace{\sigma \omega_r L_r i_{qr}}^{\Delta_{dr}} \\ V_{qr} = R_r i_{qr} + \sigma L_r \frac{di_{qr}}{dt} + \overbrace{\sigma \omega_r L_r i_{dr} + \frac{L_m}{L_s} \omega_r \varphi_{ds}}^{\Delta_{qr}} \end{cases} \tag{10}$$

where $\sigma = L_r - L_m^2/L_s$ is the machine leakage coefficient and $\omega_r = \omega_s - \omega_m$ is the slip angular frequency.

As shown in Equation (9), the stator active and reactive powers are independent from each other. In addition, the components of stator power are linearly varying with the direct and quadrature rotor currents. This powers' regulation is thus performed using PI controllers for the *d* and *q* axis components of the rotor currents.

The outer control loops for the stator active and reactive powers are tuned based on the proposed metaheuristics-based procedure. Hence, the design of PI controllers for the inner rotor current loops is detailed according to Figure 2.

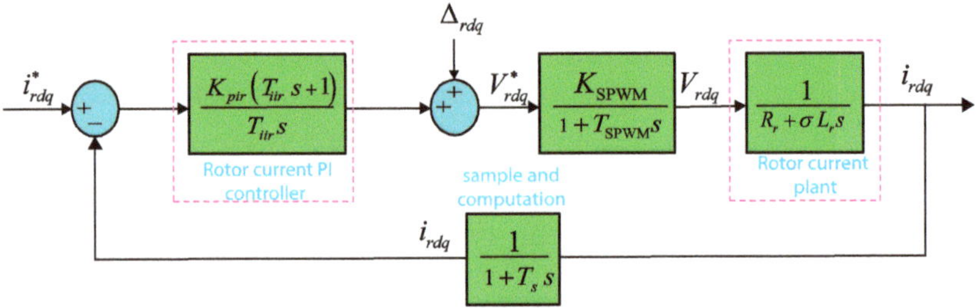

Figure 2. Proportional-Integral (PI) control scheme for the inner current loop of the RSC.

Based on a pole placement method as follows:

$$R_{PI}(s) = K_{pir}\left(1 + \frac{1}{T_{iir}s}\right) \tag{11}$$

where K_{pir} and T_{iir} are the proportional gain and time constant of the rotor PI controllers, respectively.

The transfer function that describes the relationship between the voltages and currents dynamics is given as follows [8,15]:

$$R_{\text{plant}}(s) = \frac{i_{rdq}}{V_{rdq}} = \frac{1}{R_r + \sigma L_r s} \tag{12}$$

Since the same transfer functions for the direct and quadrature rotor currents are considered, the closed-loop transfer function $R_{cli}(s)$ between the reference rotor current i^*_{rdq} and the actual one i_{rdq} is written as:

$$R_{cli}(s) = \frac{i_{rdq}}{i^*_{rdq}} = \frac{K_{pir}(T_{iir}s + 1)}{T_{iir}\sigma L_r s^2 + T_{iir}\left(R_r + K_{pir}\right)s + K_{pir}} \tag{13}$$

By equating the denominator of the closed-loop transfer function of Equation (13) under the general form of a second order system, the parameters of PI rotor current controllers can be found as follows [8,15]:

$$\begin{cases} K_{pir} = 2\xi\omega_n\sigma L_r - R_r \\ T_{iir} = \frac{2\xi}{\omega_n} - \frac{R_r}{\sigma{\omega_n}^2 L_r} \end{cases} \tag{14}$$

where ξ and ω_n are the damping coefficient and natural frequency of the desired closed-loop reference model, respectively.

For the design purpose, the discussion is made with respect to the choice of the damping coefficient ξ and natural frequency ω_n. Since the inner current loop in the cascade control scheme will have a much larger bandwidth than the one used in the outer-loop, a regulation at $1/20$ of the switching frequency $f_{sw,RSC}$ is retained. For the outer loop, a regulation between $1/20$ and $1/10$ of the inner loop bandwidth is made [5]. Then, feed forward compensations Δ_{dr} and Δ_{qr} should be added back to generate the desired rotor voltages V^*_{dr} and V^*_{qr}.

3.2. Control of the GSC

The VOC strategy is applied to control the GSC, which usually contains one outer PI control loop that regulates the DC-link voltage regardless of the magnitude and the direction of the rotor power. Two inner PI current loops that regulate the direct and quadrature gird currents are also included [16,17]. In addition, the passive damping strategy is employed to mitigate the resonance problem. The mathematical representation of GSC is built on the approximated model of the LCL-filter, which is used to tune the PI grid current controllers. Whereas, the whole transfer function of the LCL-filter with passive damping is taken into account for the stability analysis and investigation [4]. The block diagram of the grid current loops is shown in Figure 3.

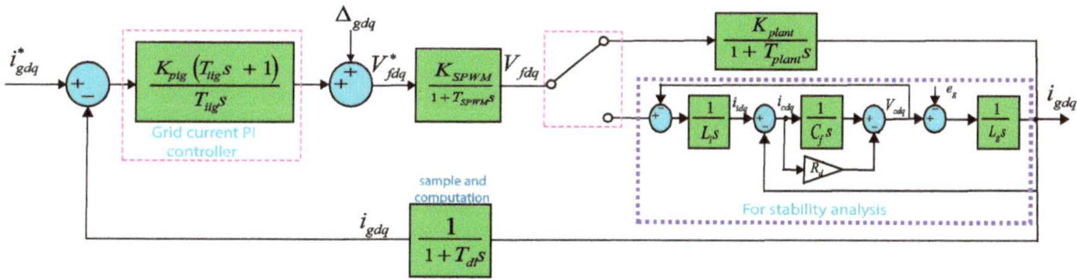

Figure 3. Design of the PI controller current loop of grid side converter.

To implement the VOC scheme, the d-axis is aligned with the grid vector voltage. Therefore, this leads to a d-axis grid voltage equal to its magnitude, and the q-axis voltage which is then equal to zero. Hence, the grid power expressions can be expressed as [16,17]:

$$\begin{cases} P_g = \frac{3}{2} e_{dg} i_{dg} \\ Q_g = -\frac{3}{2} e_{dg} i_{qg} \end{cases} \tag{15}$$

Equation (15) decides the active power and consequently the DC-link voltage is controlled via the direct current whereas the quadrature current is used to regulate the reactive power that exchanges with the grid [17]. With the VOC approach, the dynamic equations of the grid currents are rewritten as follows:

$$\begin{aligned} e_{dg} &= R_T i_{dg} - \overbrace{\omega_g L_T i_{qg}}^{\Delta_{dg}} + L_T \frac{di_{dg}}{dt} + V_{df} \\ 0 &= R_T i_{qg} + \overbrace{\omega_g L_T i_{dg}}^{\Delta_{qg}} + L_T \frac{di_{qg}}{dt} + V_{qf} \end{aligned} \tag{16}$$

By applying the same methodology of PI rotor current controllers design in the RSC, the gains of PI controllers for the grid currents dynamics are found as [15,18]:

$$\begin{cases} K_{pig} = 2\xi\omega_n L_T - R_T \\ K_{iig} = \omega_n^2 L_T \end{cases} \tag{17}$$

where $\omega_n = 847.80$ rad/sec and $\xi = 0.707$.

4. PI Controllers Tuning Problem Formulation

In the PI control framework, appropriate values of K_p and K_i gains are generally obtained by empirical methods and trials-errors based procedures [19]. These nonsystematic and challenging tasks become more difficult and time-consuming, especially for the complex and large-scale systems like the studied DFIG-based WECS. So, the idea of formulating the K_p and K_i gains' selection as an optimization problem is a promising solution. Such a control problem can be nonlinear, non-smooth or even non-convex and can be effectively solved thanks to advanced metaheuristics [10,11]. In this work, three PI controllers for the outer-loops at both RSC and GSC components are considered for the optimization process. These PI controllers for the DC-link voltage, active and reactive powers' dynamics are systematically tuned thanks to the proposed TEO metaheuristic which is described in the following. Figure 4 gives the proposed optimization-based tuning scheme of the PI controllers for the DFIG-based WECS.

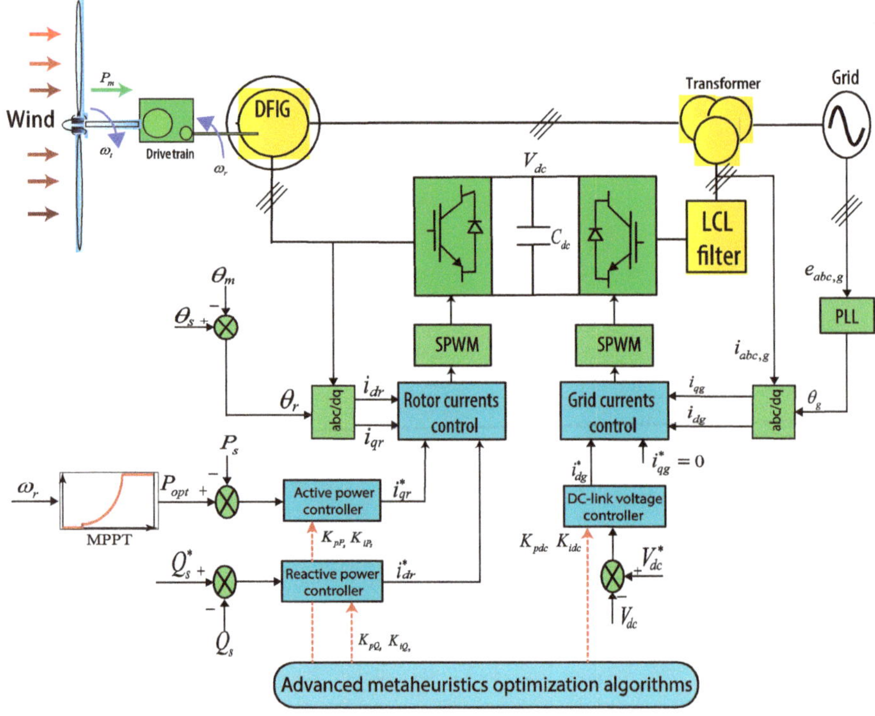

Figure 4. The proposed control algorithm applied to the DFIG system.

The decision variables of the formulated optimization problem are obviously the proportional and integral gains of PI controllers for the active/reactive powers and DC-link loops. They are specified for the optimization process as follows:

$$x = \left[K_{pP_s}, K_{iP_s}, K_{pQ_s}, K_{iQ_s}, K_{pdc}, K_{idc}\right]^T \in \mathcal{S} \subseteq \mathbb{R}^6_+ \tag{18}$$

where $\mathcal{S} = \left\{x \in \mathbb{R}^6_+, x_{low} \leq x \leq x_{up}\right\}$ denotes the bounded search space for all PI parameters.

However, the defined objective functions are minimized taking into account a scope of time-domain restrictions. These are concerning to the maximum overshoot $\delta^{\max}$, steady-state error E_{ss}, rise time t_r and/or settling time t_s of the closed-loop system step-response [20]. So, the tuning issue associating the PI controllers of the DFGI-based WECS can be expressed as follows:

$$\begin{cases} \text{Minimize } f_m(x),\ m \in \{IAE, ISE, ITSE, ITAE\} \\ x = \left[K_{pP_s}, K_{iP_s}, K_{pQ_s}, K_{iQ_s}, K_{pdc}, K_{idc}\right]^T \in \mathcal{S} \subseteq \mathbb{R}^6_+ \\ \text{subject to}: \\ g_1(x) = \delta_{P_s} - \delta^{\max}_{P_s} \leq 0 \\ g_2(x) = \delta_{Q_s} - \delta^{\max}_{Q_s} \leq 0 \\ g_3(x) = \delta_{dc} - \delta^{\max}_{dc} \leq 0 \\ K_{pj,\min} \leq K_p \leq K_{pj,\max} \\ K_{ij,\min} \leq K_i \leq K_{ij,\max},\ j \in \{P_s, Q_s, dc\} \end{cases} \tag{19}$$

where $f_m : \mathbb{R}^6_+ \to \mathbb{R}$ are the cost functions, $g_q : \mathbb{R}^6_+ \to \mathbb{R}$ are the problem's inequality constraints, δ_{dc}, δ_{P_s} and δ_{Q_s} are the overshoots of the controlled DC-voltage, active and reactive powers dynamics, respectively, $\delta^{\max}_{dc}$, $\delta^{\max}_{P_s}$ and $\delta^{\max}_{Q_s}$ denote their maximum given value.

Since the optimization problem (19) is a multi-objective type, weighted sum-based aggregation functions are used as follows to handle with the multiple costs [21]:

$$f_{IAE}(\boldsymbol{x}) = w_{P_s}\int_0^T \left|e_{P_s}(\boldsymbol{x},t)\right|dt + w_{Q_s}\int_0^T \left|e_{Q_s}(\boldsymbol{x},t)\right|dt + w_{dc}\int_0^T \left|e_{dc}(\boldsymbol{x},t)\right|dt \tag{20}$$

$$f_{ISE}(\boldsymbol{x}) = w_{P_s}\int_0^T e_{P_s}^2(\boldsymbol{x},t)dt + w_{Q_s}\int_0^T e_{Q_s}^2(\boldsymbol{x},t)dt + w_{dc}\int_0^T e_{dc}^2(\boldsymbol{x},t)dt \tag{21}$$

$$f_{ITSE}(\boldsymbol{x}) = w_{P_s}\int_0^T te_{P_s}^2(\boldsymbol{x},t)dt + w_{Q_s}\int_0^T te_{Q_s}^2(\boldsymbol{x},t)dt + w_{dc}\int_0^T te_{dc}^2(\boldsymbol{x},t)dt \tag{22}$$

$$f_{ITAE}(\boldsymbol{x}) = w_{P_s}\int_0^T t\left|e_{P_s}(\boldsymbol{x},t)\right|dt + w_{Q_s}\int_0^T t\left|e_{Q_s}(\boldsymbol{x},t)\right|dt + w_{dc}\int_0^T t\left|e_{dc}(\boldsymbol{x},t)\right|dt \tag{23}$$

where T denotes the total simulation time, $w_{P_s} > 0$, $w_{Q_s} > 0$ and $w_{dc} > 0$ are the weighting coefficients of the aggregation functions satisfying $w_{P_s} + w_{Q_s} + w_{dc} = 1$, and $e_j(.), j \in \{P_s, Q_s, dc\}$ denote the tracking errors between the plant output and the relative set-point values, i.e., $e_{P_s}(\boldsymbol{x},t) = P_s^* - P_s(\boldsymbol{x},t)$, $e_{Q_s}(\boldsymbol{x},t) = Q_s^* - Q_s(\boldsymbol{x},t)$ and $e_{dc}(\boldsymbol{x},t) = V_{dc}^* - V_{dc}(\boldsymbol{x},t)$.

The judgment matrix method is used to determine the weighting coefficients of the objective functions [22,23]. Such a method grades all objective functions based on the importance of each one. After calculating the eigenvalues of such a matrix, the weighting coefficients of functions (20)–(23) can be chosen as $w_{P_s} = 0.6370$, $w_{Q_s} = 0.2583$ and $w_{dc} = 0.1047$.

5. Thermal Exchange Optimization Algorithm

The Thermal Exchange Optimization (TEO) algorithm is a novel metaheuristic inspired by Newton's law of the cooling [3]. In this population-based metaheuristic, each agent is modeled as a cooling object. By associating another agent as a surrounding fluid, a heat transfer and thermal exchange occurs between them. The new temperature of each agent is considered as a new position of the potential solution in the search space.

In a d-dimensional search space and at the kth iteration time, each object of the population is described by its temperature $\boldsymbol{T}_i^k = \left(T_{i,1}^k, T_{i,2}^k, \ldots, T_{i,d}^k\right)$, $(i,k) \in \left[1, N_{pop}\right] \times [0, N_{iter}]$. In order to enhance the optimization performances, the TEO algorithm uses a Thermal Memory (TM) to store a number of historically best vectors as well as their related fitness. Therefore, these stored solution vectors are added to the population and the same numbers of current worst objects are deleted. After that, a growing order of solutions is retained according to their cost function values. The N_{pop} sorted objects are equally divided into two groups of environment and cooling objects as shown in Figure 5. The environment objects are denoted as $T_1^k, T_2^k, \ldots, T_{N_{pop}/2}^k$ while the cooling ones are $T_{N_{pop}/2_{+1}}^k, T_{N_{pop}/2_{+2}}^k, \ldots, T_{N_{pop}}^k$.

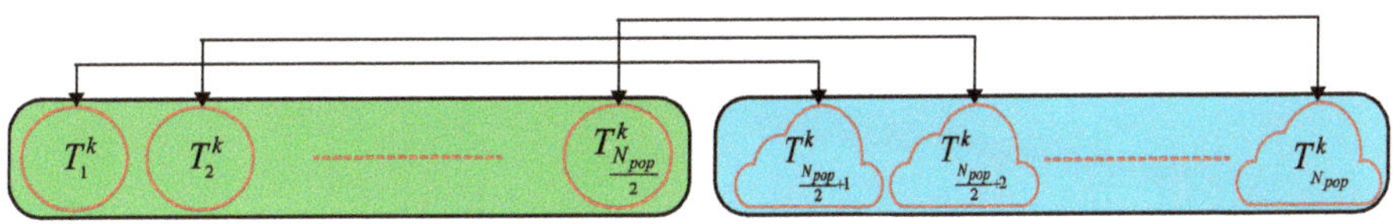

Figure 5. Pairs of environment and cooling objects.

Since the object has lower η can exchange the temperature slightly, the TEO metaheuristic proposed a similar formula to evaluate the value of η for each object as described in Equation (24):

$$\eta = \frac{\text{Cost (object)}}{\text{Cost(worst object)}} \tag{24}$$

Over the course of iterations, the value $t = k/N_{iter}$ increases as follows leading to improving the exploration mechanism [3]. In order to enhance the TEO exploration capacity, Equation (25) has been proposed to avoid the trapping in local optima and update the environmental temperature as follows:

$$T_{env,i}^{k+1} = (1 - \mathcal{U}\{0,1\}(c_1 + c_2(1-t)))T_{env,i}^{k} \tag{25}$$

where c_1 and c_2 are the controlling variables chosen as 0 or 1, $\mathcal{U}\{0,1\}$ is a uniformly random number, $T_{env,i}^{k}$ is the previous environmental temperature, $(1-t)$ is proposed to reduce the randomness by closing to the last iterations. When the decreasing of the randomness is not required, the value of c_2 is set to zero.

The new temperatures for each object, i.e., either cooling objects or environmental ones, are updated as follows:

$$T_i^{k+1} = T_{env,i}^{k} + \left(T_i^k - T_{env,i}^{k}\right)e^{-\eta t} \tag{26}$$

In the TEO formalism, another formulation is suggested to improve the exploration ability. A control parameter $0 \leq pro \leq 1$ is introduced and determines whether a component of each cooling object must be changed or not. For each object, such a parameter is compared to $Rand(i), \left(i = 1, 2, \ldots, N_{pop}\right)$, which is a random number uniformly distributed between 0 and 1. If $Rand(i) < pro$, one dimension of the ith object is chosen randomly and its value is regenerated as follows [3,24]:

$$T_{(i,j)} = T_{j,\min} + \mathcal{U}(0,1) \times \left(T_{j,\max} - T_{j,\min}\right) \tag{27}$$

where $T_{(i,j)}$ is the jth component of the ith object, $T_{j,\max}$ and $T_{j,\min}$ are the upper and lower bounds of the jth component, respectively.

Finally, the steps of the proposed TEO algorithm are summarized as follows:

Step 1. Randomly initialize the temperature for all objects $T_i^0, i = 1, 2, \ldots, N_{pop}$.
Step 2. Calculate the fitness of each search object.
Step 3. Save some T best vectors and their related cost values in the TM.
Step 4. Add the saved solutions and remove the same numbers of the worst objects.
Step 5. Arrange the objects according to their related fitness in an ascending order.
Step 6. Divide the objects into two equal groups: environment and cooling objects.
Step 7. Calculate the parameters η and t.
Step 8. Change the environment temperatures by Equation (25).
Step 9. Update the temperatures according to Equations (26) and (27).
Step 10. Check the termination criterion and repeat the iterations.

To describe the proposed metaheuristics-based tuning strategy of the PI controllers in the studied DFIG-based energy conversion system, a detailed flowchart is shown in Figure 6.

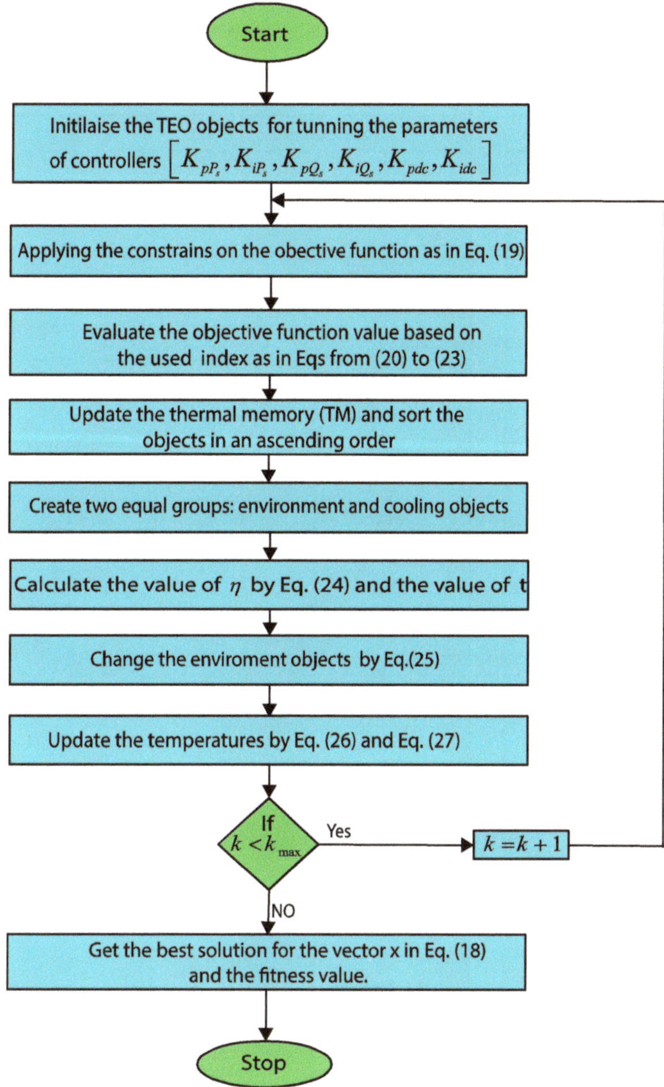

Figure 6. Flowchart of the proposed Thermal Exchange Optimization (TEO)-tuned PI controllers' parameters.

6. Simulation Results and Discussions

6.1. Execution of the Metaheuristic Algorithms

The proposed metaheuristics-based direct power control of the studied DFIG-based WECS is built using the MATLAB/Simulink environment. The respective nominal parameters of the DFIG, power converters at both sides, L-filter and LCL-filter are presented in Table 1. Specifically, this work deals with the PI controllers tuning for the active and reactive powers loops in the RSC circuit, as well as it treats with the PI controller optimization-based selection of the DC-link voltage loop in the GSC component. It is worth indicating that the inner current loops in the RSC and GSC are tuned according to the pole assignment method. Since the grid current closed-loop of the LCL-filter is unstable, the passive damping method is applied as a solution to ensure the stability of such current dynamics. To investigate the effectiveness and superiority of the proposed TEO method, the well-known global algorithms GA, PSO, HSA, WCA and GOA are considered for comparison purposes.

Table 1. Parameters for a 1.5 MW DFIG systems.

Equipment	Parameter	Value	Unit
DFIG	Rated power	1500	kW
	RMS grid line voltage	575	V
	Slip range	0.2	-
	Rated electrical frequency	50	Hz
	Stator resistance	0.023	pu
	Stator leakage inductance	0.18	pu
	Rotor resistance	0.016	pu
	Rotor leakage inductance	0.16	pu
	Magnetizing inductance	2.9	pu
Power converters	Rated power	300	kW
	Switching frequency at $f_{sw,RSC}$ and $f_{sw,GSC}$	2700	Hz
	DC-link voltage	1050	V
	DC-link capacitance	10000	µF
L-filter	L-filter grid inductance	0.1	pu
LCL-filter	LCL-filter grid side inductance	0.018	pu
	LCL-filter capacitance	0.104	pu
	LCL-filter converter side inductance	0.077	pu
	Passive damping resistor	0.124	pu

All reported algorithms are independently run 10 times. The termination criterion is set as a maximum number of iteration reached $N_{iter} = 100$ for a population size of the thermal agents equal to $N_{pop} = 50$. All the values of the used common parameters are kept equal. All algorithms have been executed on a PC computer with Core TM i5-7200U CPU and 2.5 GHz/8.00 GB RAM. The specific control parameters of the reported algorithms are listed in Table 2.

Table 2. Parameter setting of reported algorithms.

Algorithms	Parameters Setting
PSO	Cognitive and social coeffs. $c_1 = c_2 = 2$, weights $w_{\max} = 0.9$, $w_{\min} = 0.2$ [10].
GA	Roulette wheel, crossover $\mathcal{P}_{cross} = 1$, mutation $\mathcal{P}_{mut} = 0.01$ [11].
HSA	Harmony memory rate HMCR = 0.9, pitch adjusting rate PAR = 0.3 [25].
WCA	Summation number of rivers $N_{sr} = 8$ and $d_{max} = 1 \times 10^{-3}$ [18].
GOA	$c_{\max} = 1$, $c_{\min} = 0.00001$ [26].
TEO	Thermal memory $TM = 10$, $pro = 0.50$, $c_1 = 1$, $c_2 = 1$ [3].

The optimization problem (19) is minimized for various performance criteria, i.e., IAE, ISE, ITAE and Integral Time-weighted Square Error (ITSE), and under time-domain operational constraints. These performance indexes are calculated within two conditions, which are considered as follows:

- 14.3% step change in the reference of DC-link voltage at time $t = 0.5$ sec;
- step change in the reference stator reactive power at time $t = 0.8$ sec.

To find the accurate values of the decision variables $x = \left[K_{pP_s}, K_{iP_s}, K_{pQ_s}, K_{iQ_s}, K_{pdc}, K_{idc}\right]$, the bounded search domain is initially set to $0 \leq x \in \mathbb{R}_+^6 \leq 10^5$. Several runs of the algorithm are performed within this initial limitation. The results indicated that the search space could be reduced to $0 \leq x \in \mathbb{R}_+^6 \leq 400$. Therefore, the TEO algorithm explores within a smaller search domain for the next runs and more precise solution output vector is achievable.

Table 3 gives the statistical results attained by the introduced algorithms under minimizing the cost functions described in Equations (20)–(23). The worst, average, best values and the standard deviation (STD) experimental results are summarized in the Table 3, where the optimal mean value of each index is highlighted in bold and underlined. In addition, the Elapsed Time (ET) is also listed, which is defined as the time that any algorithm requires finding the best solution. It can be clearly observed that the proposed TEO produces very competitive solutions with the reported algorithms. Appendix A summarizes the obtained gains of the PI controllers for each proposed optimization method. Indeed, these tuned PI controllers' gains lead to the best transient and steady state responses of the entire reported algorithms.

Table 3. Statistical results of optimization problem (19) over 10 independent runs.

Indices		PSO	GA	HSA	WCA	GOA	TEO
IAE	Best	1.57	1.62	1.57	1.52	1.60	1.55
	Mean	1.73	1.67	1.60	1.59	2.01	**1.58**
	Worst	2.69	1.73	1.64	1.78	3.20	1.60
	STD	3.4×10^{-1}	3.2×10^{-2}	2.3×10^{-2}	7.6×10^{-2}	6.3×10^{-1}	1.9×10^{-2}
	ET (sec)	29022	28720	25480	32900	20400	24640
ISE	Best	33.17	34.86	32.64	32.66	34.06	31.45
	Mean	41.42	37.25	35.51	45.88	45.06	**33.46**
	Worst	61.80	40.53	38.78	61.24	64.50	36.79
	STD	8.27	1.92	1.80	10.77	9.08	1.64
	ET (sec)	28140	27420	19083	26900	19200	23420
ITSE	Best	17.47	17.49	18.45	17.14	17.86	17.79
	Mean	18.46	**17.96**	23.18	18.32	19.83	18.24
	Worst	20.45	18.21	26.77	19.75	28.11	19.20
	STD	0.921	0.207	3.616	8.1E-1	2.96	4.1E-1
	ET (sec)	21560	26280	18720	27540	19500	23380
ITAE	Best	0.650	0.632	0.646	0.640	0.644	0.642
	Mean	0.683	**0.650**	0.683	0.659	0.688	0.658
	Worst	0.707	0.62	0.718	0.736	0.742	0.675
	STD	0.018	0.019	0.021	0.029	0.040	0.014
	ET (sec)	20060	20480	16880	25880	20840	19680

In addition, Figure 7 shows the convergence histories of the mean objective function values of IAE, ISE, ITAE and ITSE criteria, respectively. Indeed, it is shown that the proposed TEO metaheuristic for both IAE and ISE indices outperforms the other reported methods in terms of the fastness and non-premature convergence as well as the solutions' quality. The TEO-based method for ITAE and ITSE criteria gives the best solution as a second order after the GA one. From these results, the superiority of the TEO metaheuristic is still shown in terms of exploitation and exploration capabilities for local and global searches. This further justifies the use of such a global metaheuristic to systematic and easy design of the direct power control of the DFIG-based WECS. The Box-and-Whisker plots of the mean objective function values are shown in Figure 8. From these results, one can observe that the boxplots of the TEO algorithm for all given performance criteria are generally lower and narrower than other algorithms. This confirms the high reproductivity of the TEO algorithm toward finding the optimal values of the solution.

In the remainder of the results, the obtained gains of the optimized PI controllers are used to assess the frequency-and time-domains performances of the outer-loops at the RSC and GSC power components. Moreover, comparisons with the classical pole placement [8,15], frequency response [5] symmetrical optimum [27], Ziegler–Nichols [28] and Tyreus–Luyben [29] tuning methods are made for the DC-link voltage dynamics as shown in Table 4 and Figure 9. The closed-loop response of the DC-link voltage dynamics is investigated under variable voltage profile. The DC-link voltage reference is varied up and down at various step levels. Figure 9 describes such a response around the final set-point value for different tuned PI controllers. Although, in real systems, the DC-link voltage is required to be constant. This scenario is proposed to check the capability of the proposed TEO algorithm under different circumstances. The aim is to show the difference between the classical tuning methods and the proposed optimization-based one. Referring to this result, the ITAE-based TEO algorithm can regulate the DC-link voltage dynamics with higher performance compared to the other algorithms.

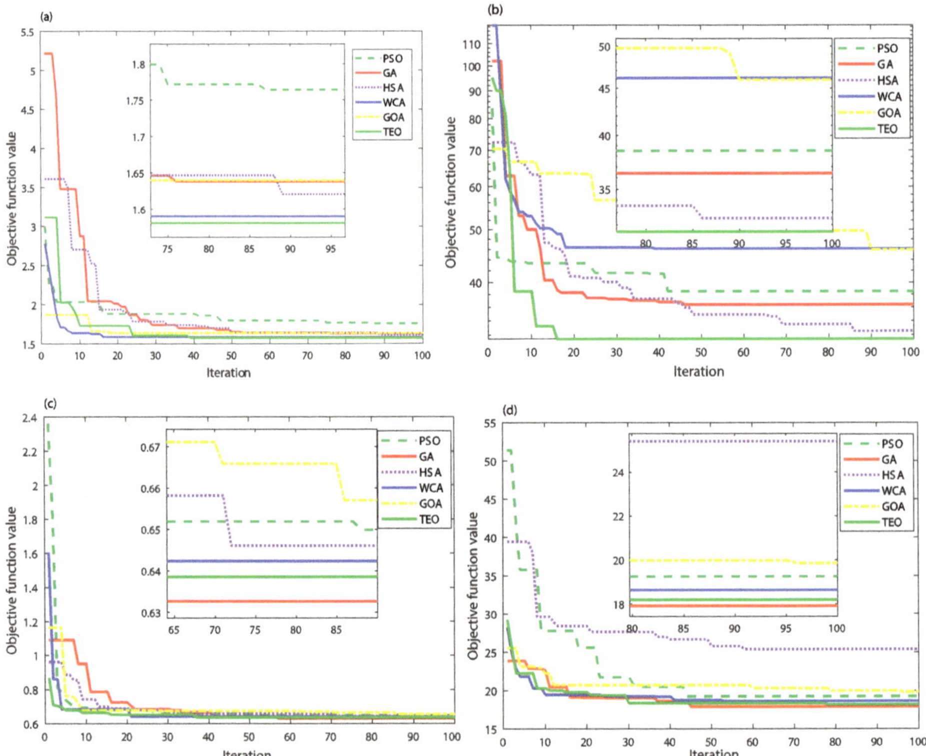

Figure 7. Convergence rates comparison: (**a**) IAE criterion; (**b**) ISE criterion; (**c**) ITAE criterion; (**d**) ITSE criterion.

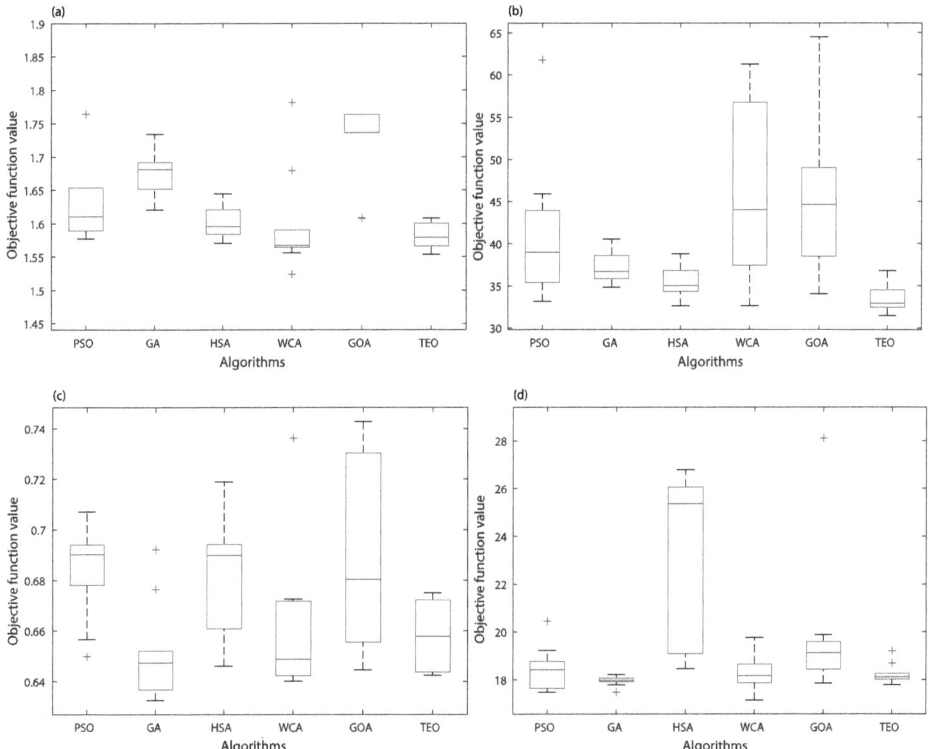

Figure 8. Box-and-Whisker plot of optimization problem (19): (**a**) IAE criterion; (**b**) ISE criterion; (**c**) ITAE criterion; (**d**) ITSE criterion.

Table 4. Time-domain performances for controlled DC-link voltage under step changes scenario.

PI Tuning Methods	Unit Step Change Response				
	t_r (sec)	t_s (sec)	t_p (sec)	δ (%)	E_{ss}
Frequency response	0.0441	4.3813	4.1382	1.168	0.3355
Pole placement	0.0163	4.1111	4.0441	2.006	0.2813
Symmetrical optimum	0.0185	4.1324	4.0459	2.768	0.3568
Ziegler–Nichols	0.0037	4.0252	4.0080	1.209	0.2843
Tyreus–Luyben	0.0060	4.0249	4.0182	0.299	0.2977
TEO	0.0029	4.0193	4.0076	2.537	0.2539

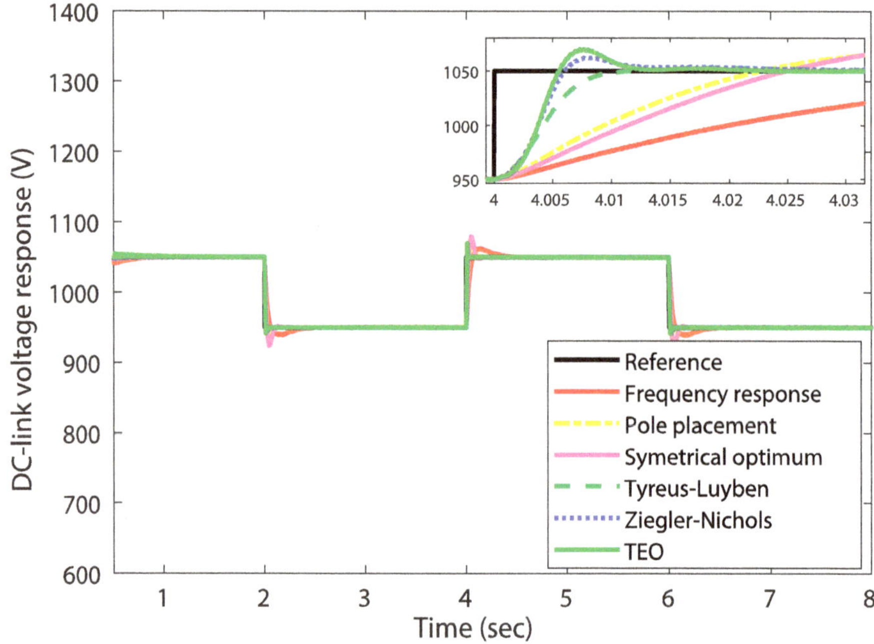

Figure 9. DC-link voltage responses against step changes for different-tuned PI controllers.

The voltage tracking is achieved with best performance in terms of response precision and fastness, i.e., the steady-state error and rise/settling time metrics are minimal as depicted in Table 4. The performance in terms of response damping is acceptable for the proposed TEO-tuned PI control as shown in Figure 9. Referring to the numerical results in Table 4, one can clearly observe that the proposed TEO algorithm showed the best rise/settling times and steady-state error except the overshoot index, which the Tyreus–Luyben method gained the minimum overshoot. Since several runs were executed to obtain the PI controllers' gains with the classical Ziegler–Nichols and Tyreus–Luyben methods, the tuning process becomes a tedious and time-consuming task. These minor degradations of the overshoot performance do not influence the effectiveness of the proposed TEO-based tuning method with respect to the systematization of the synthesis procedure, the simplicity of implementation and the superiority in other performance indices. The proposed TEO algorithm found the optimal gains of the PI controllers within a reduced computation time and under operational constraints compared to the reported trials-errors based tuning procedures which generally give local solutions for the formulated control problem.

Regarding to the grid current dynamics, the same tuning methods with adding the technical optimum instead of the symmetrical ones are used to select and adjust the PI controllers' gains. All obtained results are reported in Table 5.

Table 5. Time-domain performances for controlled quadrature grid current under a step change scenario.

PI Tuning Methods	Unit Step Change Response			
	t_r (sec)	t_p (sec)	δ (%)	E_{ss}
Frequency response	0.0171	2.0387	28.1802	0.0035
Pole placement	0.0027	2.0096	21.95	0.0045
Technical optimum	0.0146	2.0383	8.2931	0.0055
Ziegler–Nichols	3.582×10^{-4}	2.0026	10.7400	0.0051
Tyreus–Luyben	0.0020	2.0155	7.1657	0.0061

The control performance of the introduced tuning methods is investigated by supposing that there is a step change in the q-axis grid current reference at time $t = 0.2$ sec. The reference q-axis grid current is changed from zero to -0.2 pu. However, the q-axis reference grid current is usually set at zero to achieve a unity power factor. Figure 10 shows the quadrature grid current response under step changes. It can be observed that the Tyreus–Luyben method gives superior performance in comparison with the other reported methods. Here, it is important to mention that the gains of the PI controllers for the inner current loops at both RSC and GSC are selected thanks to the classical methods and not with the proposed optimization algorithms. Since the plant models of the inner loops are available, the use of the pole-placement, Tyreus–Luyben and Ziegler–Nichols methods is well adapted instead of metaheuristics-based methods which do not require models of systems to be controlled. The Bode plot of the grid current loop is presented in Figure 11.

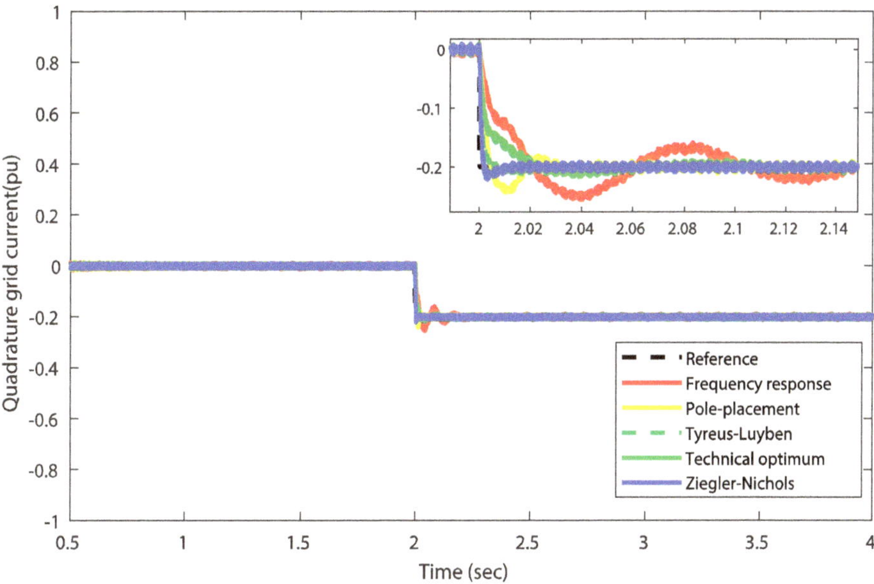

Figure 10. Quadrature grid current responses against a step change for different-tuned PI controllers.

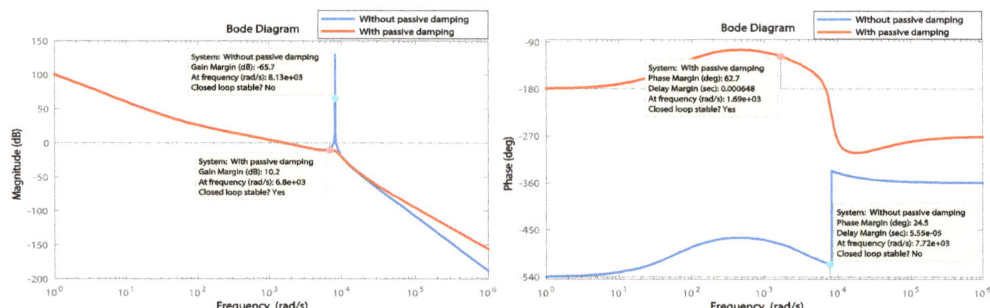

Figure 11. Bode plot of the open loop grid current loop: with and without passive damping based technical optimum tuning.

The quality of the three phase grid currents is investigated through measuring the THD of the AC grid currents. As depicted in Figure 12. It is obviously shown that the use of the LCL-filter achieves a better attenuation with a THD value about 4.05% for which the IEEE519-1992 standard limits are respected. Further analysis can be made to compare the THD of the AC grid currents in the case of using an L-filter based structure. According to the results of Figure 12, it is clear that the THD for adopting the LCL-filer is smaller than that using the L-filer type at the same value of inductance.

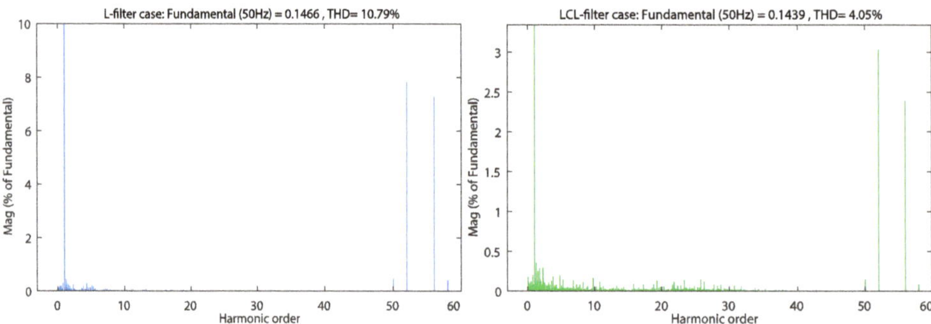

Figure 12. Harmonics spectrum of the AC grid currents.

Referring to the RSC control results, the stator active and reactive powers can be regulated by controlling the q-axis rotor current. The reference value of stator active power is generated from a Maximum Power Point Tracking (MPPT) strategy to extract the maximum power from the wind. In addition, the value of reference reactive power is set at different step levels to check the performance of the proposed PI controllers. As shown in Figure 13, both active and reactive powers track effectively their reference values with good performance in terms of speed and damping dynamics. In addition, the decoupling between the active and reactive powers is perfectly assured and the DFIG extracts the maximum available power, which is approximately about 1.5 MW. Figure 14 demonstrates the control performance of the PI controller for the reactive power loop around its final set-point value based on different tuned PI methods. The aim is to show the difference between the optimization tuning-based methods. On the other hand, both left and right sides of Figure 15 show the tracking performance of the direct and quadrature rotor currents, respectively. Such tracking dynamics are perfectly performed and improved thanks to the proposed TEO-based tuning and control method.

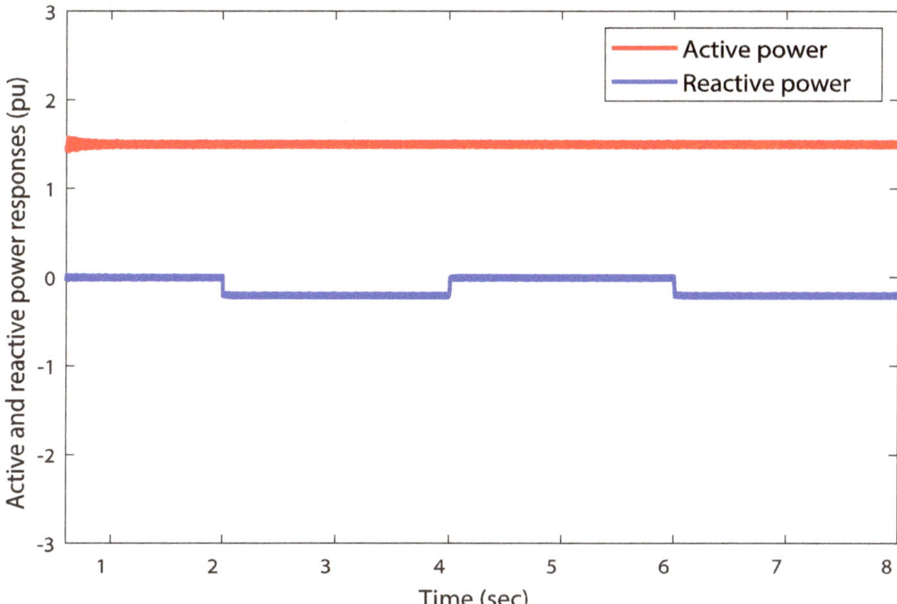

Figure 13. Time-domain variations of the TEO-tuned PI controlled active and reactive powers.

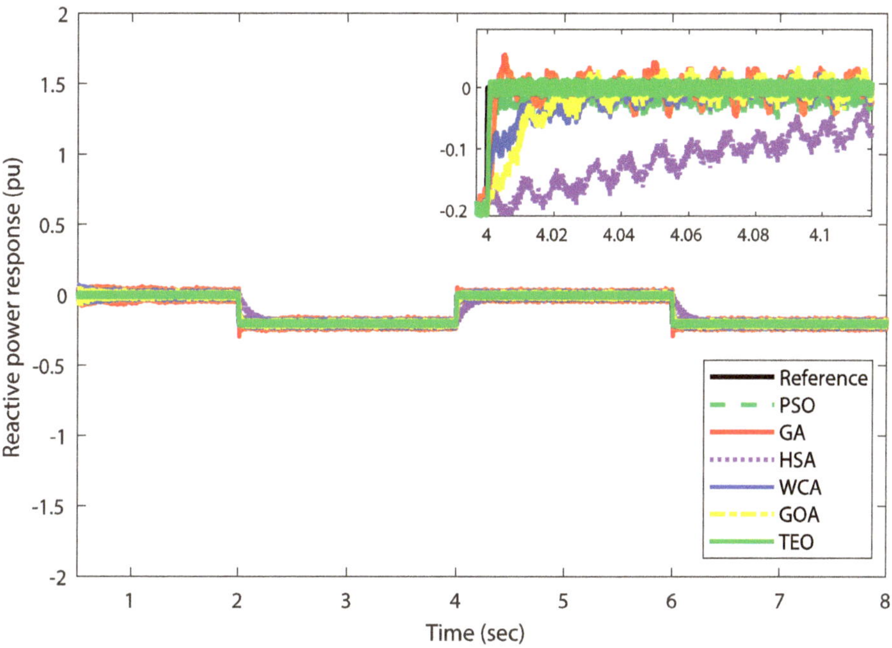

Figure 14. Performance comparison of the controlled reactive power under different PI controllers' tuning methods.

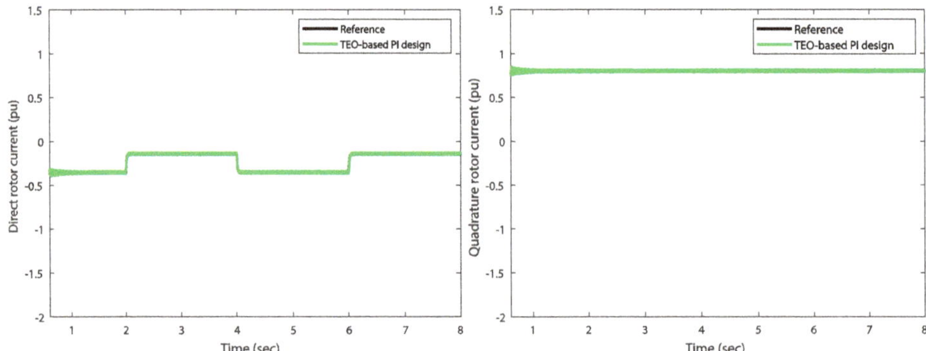

Figure 15. Time-domain variations of the TEO-based control of the direct and quadrature rotor currents.

6.2. Statistical Analysis and Comparison

In this part, the mean execution values related to the different optimization criteria are sorted to assess the best operating one according to its average objective function performance. Moreover, a statistical comparison based on the nonparametric Friedman and Bonferroni–Dunn is carried out by using these mean performances [30,31]. The average ranks for all the proposed methods based on the four performances indices are provided in Table 6. One can note that the proposed TEO metaheuristic has worthily attained the lowest average ranks compared to the remaining methods. A statistical analysis has been performed to highlight the importance of the TEO-based tuning method over other algorithms [32].

Table 6. Average Rank based statistical analysis of mean performances.

Algorithms	Indices								Average Rank
	IAE		ISE		ITSE		ITAE		
	Score	Rank	Score	Rank	Score	Rank	Score	Rank	
PSO	1.73	5	41.42	4	18.46	4	0.683	5	4.25
GA	1.67	4	37.25	3	17.96	1	0.650	1	2.25
HSA	1.60	3	35.51	2	23.18	6	0.683	4	3.75
WCA	1.59	2	45.88	6	18.32	3	0.659	3	3.75
GOA	2.01	6	44.06	5	19.83	5	0.688	6	5.5
TEO	1.58	1	33.46	1	18.24	2	0.658	2	1.5

The Friedman test for six competitor algorithms ($m = 6$) and four indices ($l = 4$) provides the computed value $\chi_F^2 = 12.14$ of the χ-distribution. The critical value of such a distribution with degrees of freedom $m - 1 = 5$ and at confidence level $\alpha = 0.05$ is equal to $\chi^2_{5,0.05} = 11.07$. Since the above computed score is greater than this statistical value, the null hypothesis is declined. Moreover, for the Iman–Davenport test [30,31], the statistic is distributed with $m - 1 = 5$ and $(m - 1) \times (l - 1) = 15$ degrees of freedom. For this test, the null hypothesis is also rejected as the calculated value $F_F = 4.64$ is greater than the critical value $F_{5,15,0.05} = 2.9$ at the same significant level of confidence. All these statistical results indicate that there are significant differences among the performances of the reported algorithms for the optimization problem (19). Thus, the proposed TEO is found to be the most effective one having the best average Friedman ranking as given in Table 6. However, a Bonferroni–Dunn post-hoc test is called to investigate whether or not the proposed TEO algorithm is significantly better than another algorithm at the above considered level of confidence [32]. The corresponding Critical Differences (CD) of the reported algorithms at the confidence levels $\alpha = 0.05$ (95% significance level) and $\alpha = 0.1$ (90% significance level) are computed as $CD_{0.05} = 3.17$ and $CD_{0.1} = 2.88$, respectively. Figure 16 illustrates the graphical representation of the Bonferroni–Dunn test considering the TEO as the control algorithm [32,33].

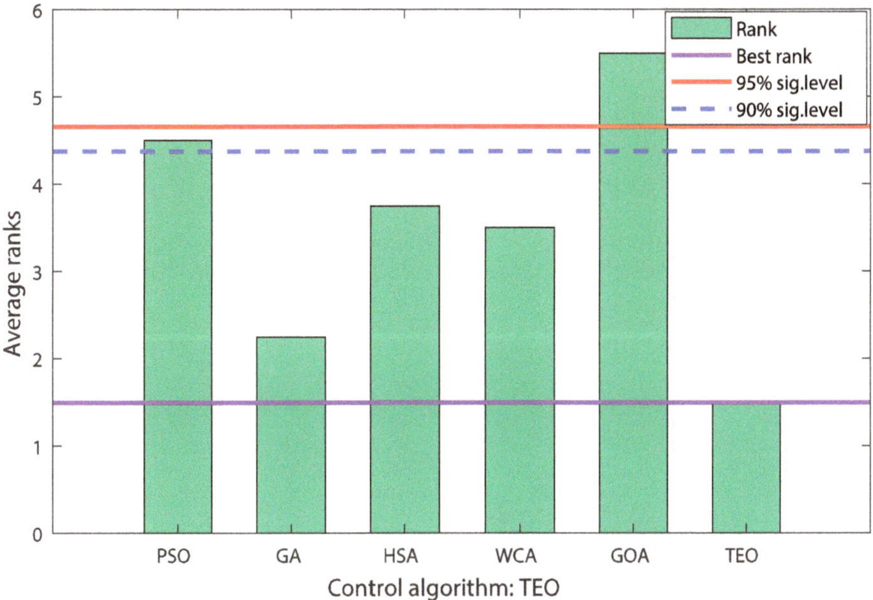

Figure 16. Graphical representation of Bonferroni–Dunn's test for problem (19).

The bar of the TEO algorithm is the lowest high among the reported algorithms and the heights of the bars corresponding to the PSO algorithm. The GOA method violates the horizontal lines of significant levels. This reveals that the TEO algorithm performs at least significantly better than these two algorithms over the solutions equality.

6.3. Computational Time Efficiency

In this subsection, the average elapsed times of Table 3 are used to assess the computational efficiency of the reported algorithms. Such an algorithmic property of resource usage is quantified by the Computational Time Efficiency (CTE) metric defined as follows [34]:

$$CTE = \frac{ET_{algorithm}}{ET_{total}}(\%) \tag{28}$$

where $ET_{algorithm}$ and ET_{total} denote the average and total elapsed times for the same index.

According to the ET measures of problem (19), the CTE for each algorithm over the reported optimization criterion is summarized in Table 7.

Table 7. The computational time efficiency (CTE) of each algorithm over the four indices.

Index	Algorithms					
	PSO	GA	HSA	WCA	GOA	TEO
IAE	18.01%	17.82%	15.81%	20.41%	12.66%	15.29%
ISE	19.52%	19.02%	13.24%	18.66%	13.32%	16.25%
ITSE	15.74%	19.19%	13.67%	20.11%	14.24%	17.07%
ITAE	16.20%	16.54%	13.63%	20.90%	16.83%	15.89%

In terms of computational efficiency, it can be observed from these results that the proposed TEO algorithm attained the second rank for IAE and ITAE criteria and the third and fourth ranks for ISE and ITSE criteria, respectively. Roughly, the HSA algorithm presents the best computational efficiency in terms of time resource usage. The GOA metaheuristic achieved the best average elapsed time as shown in Figure 17.

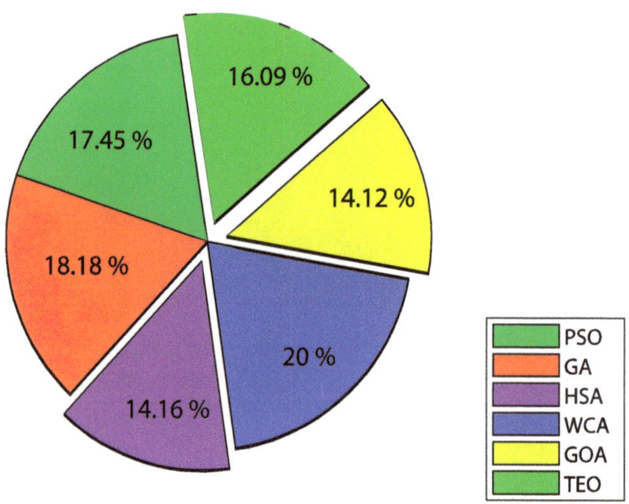

Figure 17. Average elapsed time over all criteria for the introduced algorithms.

6.4. Sensitivity Analysis

In the TEO formalism, the parameters c_1 and c_2 are introduced to reduce the randomness of the algorithm over the course of iterations and to improve the exploration capabilities. The size of thermal memory TM is considered as a predefined parameter, which can enhance the performances of the algorithm without increasing the computational cost [3,24]. However, *pro* is a user parameter that improves the global search capacity. For illustration purposes, the resolution of problem (19) is tested by changing these above main parameters. As shown in Table 8, the case with the parameters set $(c_1 = 1, c_2 = 1)$ shows the superiority of the optimization results compared to other scenarios. In addition, it can be seen from Table 9 that the user parameter $TM = 10$ leads to obtain the best solutions. Finally, Table 10 investigates the effect of the parameter *pro* on the algorithm performance. The convergence of the TEO algorithm is always guaranteed under all these control parameters variations which shows the insensitivity of the proposed physics-inspired metaheuristic.

Table 8. Comparison under different values of c_1 and c_2: IAE criterion.

Optimization Scenario	TM=10, pro=0.5 and N_{iter}=100			
	Worst	Mean	Best	STD
$c_1 = 0, c_2 = 0$	1.669	1.652	1.638	1.9×10^{-2}
$c_1 = 1, c_2 = 0$	1.664	1.640	1.622	1.4×10^{-2}
$c_1 = 0, c_2 = 1$	1.654	1.640	1.631	7.5×10^{-3}
$c_1 = 1, c_2 = 1$	1.608	1.582	1.553	1.9×10^{-2}

Table 9. Comparison under different values of TM: IAE criterion.

Optimization Scenario	c_1=1, c_2=1, pro=0.5 and N_{iter}=100			
	Worst	Mean	Best	STD
$TM = 4$	1.699	1.694	1.689	3.5×10^{-3}
$TM = 8$	1.654	1.640	1.631	8.4×10^{-2}
$TM = 10$	1.608	1.582	1.553	1.9×10^{-2}

Table 10. Comparison under different values of pro: IAE criterion.

Optimization Scenario	c_1=1, c_2=1, TM=10 and N_{iter}=100			
	Worst	Mean	Best	STD
$pro = 0.20$	1.693	1.654	1.633	1.7×10^{-2}
$pro = 0.35$	1.678	1.617	1.575	5.2×10^{-2}
$pro = 0.50$	1.608	1.582	1.553	1.9×10^{-2}

7. Conclusions

This paper proposes an intelligent metaheuristics-based design procedure to tune the outer-loop PI controllers for the direct power control scheme of DFIG-based energy conversion systems. In this study, only outer PI loops in the RSC and GSC circuits are optimized for the active and reactive powers and DC-link voltage dynamics. Since the reported classical techniques of PI controller tuning are tedious, time-consuming and not systematic, a TEO-based approach has been proposed and successfully implemented. The PI controllers tuning problem is firstly formulated as a constrained optimization program under nonlinear and non-smooth operational constraints. The introduced TEO algorithm is then employed separately to minimize several time-domain performance criteria such as IAE, ISE, ITAE and ITSE indices as objective functions. The proposed TEO-tuned PI controllers methodology improves the performance and robustness of the controlled DFIG-based energy converter in terms of rising time, settling time, and steady state indices. The classical trials-errors based methods of PI controllers tuning are no longer used and the design time is further reduced. To evaluate the performance superiority of the proposed TEO-based approach in finding the global minimum value of the objective function for various performance indices, a comparison study with the PSO, GA, HSA, GSO and WCA is performed. The demonstrative results exhibit that the proposed TEO gives very complete results in terms of global search capabilities, robustness and non-premature convergence. Finally, the statistical analysis is achieved by using the Friedman's rank and Bonferroni-Dunn's test. The corresponding results show that the proposed TEO-based method is a promising alternative approach for controlling the DFIG system by systematically tuning the unknown PI controllers' parameters efficiently.

Author Contributions: Conceptualization, M.M.A. and S.B.; Methodology, S.B.; Software, M.M.A.; Validation, M.M.A.; Formal analysis, S.B.; Investigation, S.B.; Resources, M.M.A. and S.B.; Writing—original draft preparation, M.M.A.; Writing—review and editing, S.B.; Visualization, S.B.; Supervision, S.B.

Funding: This research received no external funding.

Acknowledgments: The authors would like to express their sincere appreciation to the anonymous reviewers for their valuable and constructive comments, which greatly improve the quality of this paper.

Conflicts of Interest: The authors declare no conflict of interest.

Abbreviations

The following abbreviations are used in this manuscript:

CD	Critical Difference
CTE	Computational Time Efficiency
ET	Elapsed Time
DFIGs	Doubly Fed Induction Generators
HSA	Harmony Search Algorithm
IAE	Integral Absolute Error
IGBTs	Insulated-Gate Bipolar Transistors
ISE	Integral Square Error
ITAE	Integral Time-weighted Absolute Error
ITSE	Integral Time-weighted Square Error
GA	Genetic Algorithm
GOA	Grasshopper Optimization Algorithm
GSC	Grid Side Converter
MPPT	Maximum Power Point Tracking
PI	Proportional-Integral
PSO	Particle Swarm Optimization
RSC	Rotor Side Converter
SFO	Stator Flux Orientation
SPWM	Sinusoidal Pulse Width Modulation
STD	Standard Deviation
TEO	Thermal Exchange Optimization
THD	Total Harmonic Distortion
TM	Thermal Memory
VOC	Voltage Oriented Control
WECS	Wind Energy Converter System
WCA	Water Cycle Algorithm
WT	Wind Turbine
WTs	Wind Turbines

Notations

C_{dc}	DC-link capacitance
C_f	Filter capacitance
C_p	Power conversion coefficient
$e_{g(d,q)}$	*d*-*q* axis grid voltages
$i_{g(d,q)}$	*d*-*q* axis grid currents
$i_{r(d,q)}$	*d*-*q* axis rotor currents
$i_{s(d,q)}$	*d*-*q* axis stator currents
L_g	Filter grid side inductance
L_i	Filter converter side inductance
L_m	Magnetizing inductance
L_r, L_s	Rotor and stator inductances
L_T	Filter total inductance
P_g, P_s	Grid and stator active powers

P_m	Mechanical turbine power
Q_g, Q_s	Grid and stator reactive powers
R	Turbine radius
R_d	Filter damping resistance
R_r, R_s	Rotor and stator resistances
V_{dc}	DC-link voltage
$V_{f(d,q)}$	*d-q* axis grid converter voltage sides
$V_{r\ (d,q)}$	*d-q* axis rotor converter voltage sides
$V_{s\ (d,q)}$	*d-q* axis stator voltages
V_w	Wind speed
λ	Tip speed ratio
ω_r, ω_s	Rotor and stator angular frequencies
Ω_t	Mechanical rotational speed
ρ	Air density
$\varphi_{dr}, \varphi_{qr}$	*d-q* axis rotor fluxes
$\varphi_{ds}, \varphi_{qs}$	*d-q* axis stator fluxes

Appendix A. Decision Variables of Problem (19) Relative to the Optimization Mean Case

The following gains of the optimized PI controllers (Table A1), as decision variables of problem (19), are used to perform all numerical simulations of the paper.

Table A1. Comparative results of the PI controllers' coefficients tuning.

Indices	Algorithms	PI Controllers' Gains					
		K_{pP_s}	K_{iP_s}	K_{pQ_s}	K_{iQ_s}	K_{pdc}	K_{idc}
IAE	PSO	14.16	139.37	376.43	381.82	9	31.32
	GA	16.92	394.62	18.22	147.46	24.36	62.17
	HSA	7.89	83.96	151.27	227.72	28.03	86.11
	WCA	2.28	329.93	122.55	400	26.11	41.53
	GOA	16.91	394.62	18.22	147.47	26.36	59.85
	TEO	4.23	58.43	10.37	15.73	89.10	192.71
ISE	PSO	11.23	207.70	169.31	353	69.65	272.40
	GA	2.25	313.46	187.62	398.82	79.95	345.49
	HSA	1×10^{-5}	1.87	118.84	360.16	33.29	39.65
	WCA	12.88	143.34	360.15	400	191.11	293
	GOA	12.42	388.97	1.93	371.52	148.46	220.78
	TEO	35.94	165.25	15.44	71.59	1.48	3
ITSE	PSO	0.1	1×10^{-5}	238.93	351.81	0.99	25.03
	GA	1×10^{-5}	1×10^{-5}	7.27	214.42	24.34	62.20
	HSA	303.71	320.52	1.03	6.25	27.66	45.22
	WCA	4.13	248.98	10.79	24.45	27.89	79.04
	GOA	9.82	108.24	333.25	27.36	131.74	205.91
	TEO	8.56	5.04	67.72	69.44	10.71	31.96
ITAE	PSO	250.16	289.53	346.50	142.34	73.74	238.53
	GA	1×10^{-5}	1×10^{-5}	45.02	85.22	130.57	26.43
	HSA	8.16	4.83	323.56	384.19	33.33	47.41
	WCA	9.54	63.73	329.42	395.36	143.52	400
	GOA	13.71	399.92	302.9	265.4	97.91	196
	TEO	4.8×10^{-6}	11.28	1.6×10^{-6}	8.3×10^{-6}	10.24	68.66

References

1. Sahu, K.B. Wind Energy Developments and Policies in China: A Short Review. *Renew. Sustain. Energy Rev.* **2018**, *81*, 1393–1405. [CrossRef]
2. Belmokhtar, K.; Doumbia, M.L.; Agbossou, K. Modelling and Fuzzy Logic Control of DFIG Based Wind Energy Conversion Systems. In Proceedings of the IEEE International Symposium on Industrial Electronics (ISIE), Hangzhou, China, 28–31 May 2012; pp. 1888–1893.
3. Hu, J.; Nian, H.; Xu, H.; He, Y. Dynamic Modeling and Improved Control of DFIG under Distorted Grid Voltage Conditions. *IEEE Trans. Energy Convers.* **2011**, *26*, 163–175. [CrossRef]
4. Liserre, M.; Blaabjerg, F.; Hansen, S. Design and Control of an LCL-Filter-based Three-Phase Active Rectifier. *IEEE Trans. Ind. Appl.* **2005**, *50*, 1281–1291. [CrossRef]
5. Zhou, D.; Blaabjerg, F. Bandwidth Oriented Proportional-Integral Controller Design for Back-To-Back Power Converters in DFIG Wind Turbine System. *IET Renew. Power Gen.* **2017**, *11*, 941–951. [CrossRef]
6. Hamane, B.; Doumbia, M.L.; Bouhamida, M.; Benghanem, M. Direct Active and Reactive Power Control of DFIG based WECS using PI and Sliding Mode Controllers. In Proceedings of the 40th Annual Conference of the IEEE Industrial Electronics Society, Dallas, TX, USA, 29 October–1 November 2014; pp. 2050–2055.
7. Hamane, B.; Doumbia, M.L.; Bouhamida, M.; Draou, A.; Chaoui, H.; Benghanem, M. Comparative Study of PI, RST, Sliding Mode and Fuzzy Supervisory Controllers for DFIG based Wind Energy Conversion System. *Int. J. Renew. Energy Res.* **2015**, *5*, 1174–1185.
8. Hamane, B.; Doumbia, M.L.; Bouhamida, M.; Draou, A.; Chaoui, H.; Benghanem, M. Modeling and Control of a Wind Energy Conversion System Based on DFIG Driven by a Matrix Converter. In Proceedings of the Eleventh International Conference on Ecological Vehicles and Renewable Energies, Monte Carlo, Monaco, 6–8 April 2016; pp. 1–8.
9. Tanvir, A.A.; Merabet, A.; Beguenane, R. Real-Time Control of Active and Reactive Power for Doubly Fed Induction Generator (DFIG)-Based Wind Energy Conversion System. *Energies* **2015**, *8*, 10389–10408. [CrossRef]
10. Bekakra, Y.; Attous, D.B. Optimal Tuning of PI Controller Using PSO Optimization for Indirect Power Control for DFIG Based Wind Turbine with MPPT. *Int. J. Syst. Assur. Eng. Manag.* **2014**, *5*, 219–229. [CrossRef]
11. Vieira, J.P.A.; Nunes, M.N.A.; Bezerra, U.H. Design of Optimal PI Controllers for Doubly Fed Induction Generators in Wind Turbines Using Genetic Algorithm. In Proceedings of the IEEE Power and Energy Society General Meeting—Conversion and Delivery of Electrical Energy in the 21st Century, Pittsburgh, PA, USA, 20–24 July 2008; pp. 1–7.
12. Asgharnia, A.; Shahnazi, R.; Jamali, A. Performance and Robustness of Optimal Fractional Fuzzy PID Controllers for Pitch Control of a Wind Turbine Using Chaotic Optimization Algorithms. *ISA Trans.* **2018**, *79*, 27–44. [CrossRef]
13. Assareh, E.I.; Poultangari, I.; Tandis, E.; Nedaei, M. Optimizing the Wind Power Generation in Low Wind Speed Areas Using an Advanced Hybrid RBF Neural Network Coupled with the HGA-GSA Optimization Method. *J. Mech. Sci. Technol.* **2016**, *30*, 4735–4745. [CrossRef]
14. Kaveh, A.; Dadras, A. A Novel Meta-Heuristic Optimization Algorithm: Thermal Exchange Optimization. *Adv. Eng. Softw.* **2017**, *110*, 69–84. [CrossRef]
15. Gagra, S.K.; Mishra, S.; Sing, M. Performance Analysis of Grid Integrated Doubly Fed Induction Generator for a Small Hydropower Plant. *Int. J. Renew. Energy Res.* **2018**, *8*, 2310–2323.
16. Qiao, W. Dynamic Modeling and Control of Doubly Fed Induction Generators Driven by Wind Turbines. In Proceedings of the IEEE/PES Power Systems Conference and Exposition, Seattle, DC, USA, 15–18 March 2009; pp. 1–8.
17. Xu, D.; Blaabjerg, F.; Chen, W.; Zhu, N. *Advanced Control of Doubly Fed Induction Generator for Wind Power Systems*, 1st ed.; John Wiley & Sons: New York, NY, USA, 2018.
18. Hato, M.M.; Bouallègue, S.; Ayadi, M. Water cycle algorithm-tuned PI control of a doubly fed induction generator for wind energy conversion. In Proceedings of the 9th International Renewable Energy Congress (IREC), Hammamet, Tunisia, 20–22 March 2018; pp. 1–6.
19. Åström, A.K.; Hägglund, T. *Advanced PID Control*; ISA-Instrumentation, Systems and Automation Society: Research Triangle Park, NC, USA, 2006.

20. Bouallègue, S.; Haggège, J.; Ayadi, M.; Benrejeb, M. PID-Type Fuzzy Logic Controller Tuning based on Particle Swarm Optimization. *Eng. Appl. Artif. Intel.* **2012**, *25*, 484–493. [CrossRef]
21. Qu, B.Y.; Zhu, Y.S.; Jiao, Y.S.; Wu, M.Y.; Suganthan, P.N.; Liang, J.J. A Survey on Multi-objective Evolutionary Algorithms for the Solution of the Environmental/Economic Dispatch Problems. *Swarm Evol. Comput.* **2018**, *8*, 1–11. [CrossRef]
22. Gong, L.; Cao, W.; Liu, K.; Zhao, J.; Li, X. Spatial and Temporal Optimization Strategy for Plug-in Electric Vehicle Charging to Mitigate Impacts on Distribution Network. *Energies* **2018**, *11*, 1373. [CrossRef]
23. Zhang, X.; Yang, J.; Wang, W.; Jing, T.; Zhang, M. Optimal Operation Analysis of The Distribution Network Comprising a Micro Energy Grid based on an Improved Grey Wolf Optimization Algorithm. *Appl. Sci.* **2018**, *8*, 923. [CrossRef]
24. Kaveh, A.; Ghazaan, M.I. Enhanced Colliding Bodies Optimization for Design Problems with Continuous and Discrete Variables. *Adv. Eng. Softw.* **2014**, *77*, 66–75. [CrossRef]
25. Geem, Z.W.; Kim, J.H.; Loganathan, G.V. A New Heuristic Optimization Algorithm: Harmony Search. *Sage J.* **2012**, *76*, 60–68.
26. Saremi, S.; Mirjalili, S.; Lewis, A. Grasshopper Optimization Algorithm: Theory and Application. *Adv. Eng. Softw.* **2017**, *105*, 30–47. [CrossRef]
27. Blasko, V.; Kaura, V. A Novel Control to Actively Damp Resonance in Input LC Filter of a Three-Phase Voltage Source Converter. *IEEE Trans. Ind. Appl.* **1997**, *33*, 542–550. [CrossRef]
28. Bingi, K.; Ibrahim, R.; Karsiti, M.N.; Chung, T.D.; Hassan, S.M. Optimal PID control of pH neutralization plant. In Proceedings of the 2nd IEEE International Symposium on Robotics and Manufacturing Automation, Ipoh, Malaysia, 25–27 September 2016; pp. 1–6.
29. Mishra, K.P.; Kumar, V.; Rana, K.P.S. Stiction combating intelligent controller tuning a comparative study. In Proceedings of the International Conference on Advances in Computer Engineering and Application, Ghaziabad, India, 19–20 March 2015; pp. 534–541.
30. Raghuwanshi, B.S.; Shukla, S. Generalized Class-Specific Kernelized Extreme Learning Machine for Multiclass Imbalanced Learning. *Expert. Syst.* **2019**, *121*, 244–255. [CrossRef]
31. Zeng, F.; Zhang, W.; Zhang, S.; Zheng, N. Re-Kissme: A Robust Resampling Scheme for Distance Metric Learning in the Presence of Label Noise. *Neurocomputing* **2019**, *330*, 138–150. [CrossRef]
32. Sadhu, A.K.; Konar, A.; Bhattacharjee, A.T.; Das, S. Synergism of Firefly Algorithm and Q-Learning for Robot Arm Path Planning. *Swarm Evol. Comput.* **2018**, *43*, 50–68. [CrossRef]
33. Liu, Y.; Chen, S.; Guan, B.; Xu, P. Layout Optimization of Large-Scale Oil–Gas Gathering System based on Combined Optimization Strategy. *Neurocomputing* **2019**, *332*, 159–183. [CrossRef]
34. Sababha, M.; Zohdy, M.; Kafafy, M. The Enhanced Firefly Algorithm Based on Modified Exploitation and Exploration Mechanism. *Electronics* **2018**, *7*, 132. [CrossRef]

© 2019 by the authors. Licensee MDPI, Basel, Switzerland. This article is an open access article distributed under the terms and conditions of the Creative Commons Attribution (CC BY) license (http://creativecommons.org/licenses/by/4.0/).

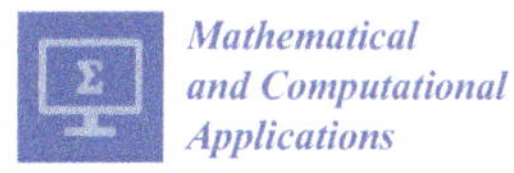

Article

Adaptive Control of Resistance Spot Welding Based on a Dynamic Resistance Model

Ziyad Kas and Manohar Das *

Department of Electrical and Computer Engineering, Oakland University, Rochester, MI 48309, USA; z_kas@hotmail.com
* Correspondence: das@oakland.edu

Received: 16 August 2019; Accepted: 27 September 2019; Published: 28 September 2019 

Abstract: Resistance spot welding is a process commonly used for joining a stack of two or three metal sheets at desired spots. Such welds are accomplished by holding the metallic workpieces together by applying pressure through the tips of a pair of electrodes and then passing a strong electric current for a short duration. This kind of welding process often suffers from two common drawbacks, namely, inconsistent weld quality and inadequate nugget size. In order to address these problems, a new theoretical approach of controlling resistance spot welding processes is proposed in this paper. The proposed controller is based on a simplified dynamical model of the resistance spot welding process and employs the principle of adaptive one-step-ahead control. It is essentially an adaptive tracking controller that estimates the unknown process parameters and adjusts the welding voltage continuously to make sure that the nugget resistance tracks a desired reference resistance profile. The modeling and controller design methodologies are discussed in detail. Also, the results of a simulation study to evaluate the performance of the proposed controller are presented. The proposed control scheme is expected to reduce energy consumption and produce consistent welds.

Keywords: resistance spot welding; dynamic resistance model; adaptive control; energy savings

---

1. Introduction

Resistance spot welding (RSW) is an electrothermal process in which contacting metal surfaces are joined by heat. The metal surfaces are held together under pressure exerted by two electrodes. The heat needed to create the weld is generated by applying a strong electric current through the electrodes and the workpieces, as shown in Figure 1. The welding process is related to the metallurgy of the materials involved in welding, including the base metal and the electrodes [1]. The flow of a strong electric current through the metal sheets causes heating due to the resistance of the joining surfaces and the sheets. Most of the heating is concentrated near the faying surface, since the contact resistance is very high compared to the bulk resistance of the sheets, causing melting and formation of a weld nugget. Depending on the thickness and type of material, welding current ranges from 1000 to 20,000 amperes, or even higher, while the voltage typically lies between 1 and 30 volts [2].

A resistance spot welding cycle consists of three main stages as follows:

Stage 1: Squeeze time, during which pressure is applied by the electrodes to squeeze the workpieces.
Stage 2: Welding time, during which a high current is applied, causing melting and formation of a nugget.
Stage 3: Hold time, during which pressure is maintained after the welding current is ceased, to allow cooling of the nugget and prevent cracks.

The resistance spot welding process is used in many different industries, including automotive, aerospace, railway, military, and industrial manufacturing. It is the most popular welding technology used by the automotive industry to weld various sheet metals to form the chassis and body of a vehicle. About 4000–6000 spot welds are used to manufacture a typical automotive vehicle today. Considering a worldwide annual production volume of 80 million automotive vehicles, an energy efficient RSW controller can result in significant energy savings and reduce carbon footprint accordingly.

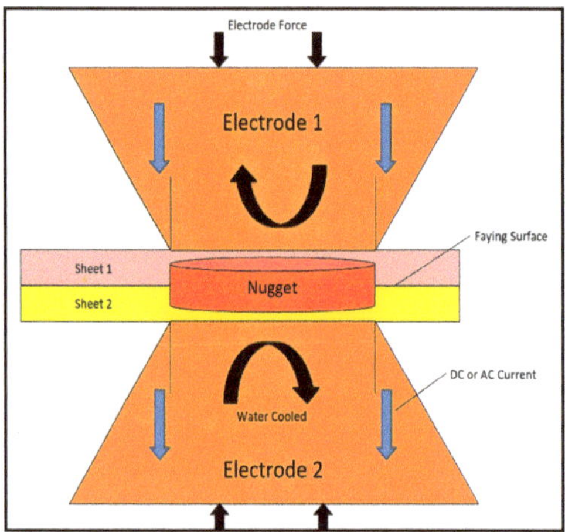

Figure 1. Resistance spot welding system.

RSW welding often suffers from two common drawbacks, namely, inconsistent weld quality and inadequate nugget size. In order to address these problems, a number of remedies have been proposed. These include monitoring and control of welding parameters to improve weld quality. Some of the conventional RSW control techniques proposed to date include proportional-integral (PI) [3], proportional-derivative (PD) [4], proportional-integral-derivative (PID) [5], fuzzy [6–8], neural networks [9,10], or a combination of fuzzy and neural networks [11]. One of the main drawbacks of these techniques stems from the fact they do not take into consideration the thermal dynamics of an RSW process. Also, most of these systems do not take into account any changes in process parameters. Some of the recently proposed RSW control methods address the above drawbacks by proposing either model-based or adaptive controllers. These include a power reference-based controller [12], intelligent hysteresis and PI controllers [13], and an adaptive power and current controller to reduce shunting effects [14].

As a remedy to some of the shortcomings of the existing methodologies, a new RSW control design method based on using dynamic resistance as a feedback signal is proposed in this paper. Although the idea of using dynamic resistance signature for monitoring quality of RSW welds has been investigated by a number of researchers over past three decades [15–20], the idea of using it as a feedback signal for controlling an RSW process is relatively new. Although this idea was introduced by a number of researchers, such as [17,21], no methods of actually doing it were presented. However, a method of doing it using model predictive control (MPC) was proposed in [22]. It introduces the idea of a dynamic resistance tracking controller, which is conceptually similar to the controller proposed in this paper. However, the RSW resistance model developed there is an empirical dynamical resistance model consisting of three first order exponential curves to emulate the characteristics of the resistance of the workpiece during the heating, melting, and cooling phases, respectively. It proposes a model predictive controller (MPC) based on the above simplistic empirical model, but judging from the results presented, it does not seem to perform well.

As mentioned earlier, an RSW controller design based on using dynamic resistance as a feedback signal is proposed in this paper. It is based on a dynamic resistance model of the RSW process, which is slightly different from the thermal RSW process models investigated in [23,24]. The main drawback of using a thermal model for RSW control has to do with the difficulty of directly monitoring the nugget temperature, which is why an indirect monitoring scheme based on measurement of the dynamic resistance was suggested in [23,24]. The above difficulty is avoided here by developing and utilizing a dynamic resistance model of the RSW process. The development of such a model starts from a simplified heat balance model of an RSW process proposed in [23–25], which is then converted into a dynamical resistance model using a nonlinear relationship between temperature and resistance. The resulting model governs the variation of workpiece resistance during welding time. Next, an adaptive one-step-ahead (AOSA) controller is designed based on the above model. The proposed adaptive controller estimates the model parameters and generates a control signal based on the estimated parameters, which enable the workpiece resistance to follow a desired reference resistance profile. Simulation results show that an AOSA controller is capable of tracking a reference resistance profile when the welding parameters are unknown, as well as reducing the energy required to make a weld.

The organization of this paper is as follows. Section 2 presents modeling of an RSW nugget formation process. The development of a dynamical resistance model and its validation are discussed in Section 3. The design of an adaptive controller is discussed in Section 4. Section 5 presents the results of some simulation studies, and finally, some concluding remarks are provided in Section 6.

2. Modeling of an RSW Nugget Formation Process

This section presents a dynamical resistance model of the RSW process. First, a simplified heat balance model of the process is described, and then it is converted to a dynamical resistance model.

Electro-Thermal Dynamical Model of an RSW Nugget Formation Process

Following the footsteps of [23–25], the development of an electro-thermal model is started from a simplified RSW nugget model, shown in Figure 2 below. The heat balance equations for this model can be developed as follows.

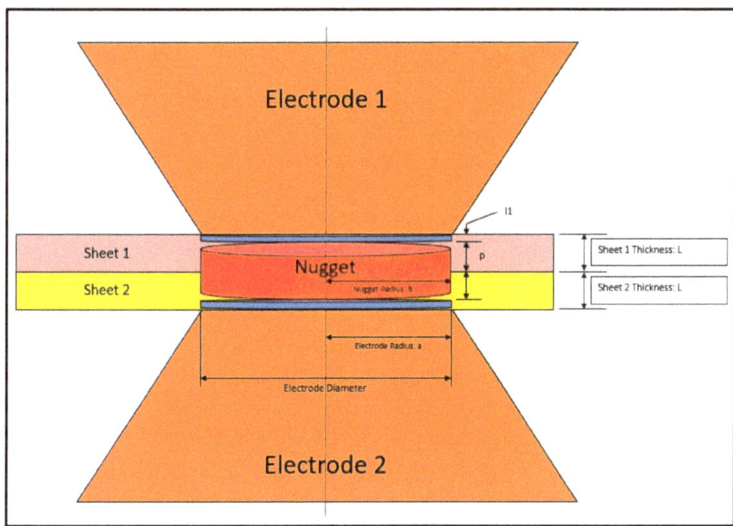

Figure 2. A simplified thermal model of nugget formation.

The total heat generation rate, $\dot{Q}_g(t)$, is given by:

$$\dot{Q}_g(t) = \frac{V^2(t)}{R(t)} \tag{1}$$

where $V(t)$ denotes the welding voltage and $R(t)$ is the total workpiece resistance, which can be described by:

$$R(t) = R_w(t) + R_c(t) + R_e(t). \tag{2}$$

Here, $R_w(t)$ denotes the bulk resistance of the workpieces, $R_c(t)$ represents the total contact resistance, and $R_e(t)$ denotes the electrode resistance. Since R_e is very small compared to the other two components, it can be neglected in (2). It may be noted that for a two-stack workpiece, the total resistance can also be rewritten as:

$$R(t) \approx 2R(t)_{electrode-sheet} + R(t)_{sheet-sheet} + R_w(t) \tag{3}$$

The heat of fusion required for nugget formation is given by:

$$H_f = HV_n, \tag{4}$$

where H denotes the heat of fusion per unit volume and V_n denotes the nugget volume, which is given by:

$$V_n = 2\pi a^2 p, \tag{5}$$

where p and a are the penetration radius and nugget radius, respectively. Substitution of (5) into (4) and normalization over the weld duration, Δt, yields:

$$c_1 = \frac{H_f}{\Delta t} = 2H\pi a^2 p. \tag{6}$$

Neglecting the heat loss to the surroundings and the electrodes, the heat required to raise the temperature of the nugget by $d\theta(t)$ is given by:

$$dQ_T(t) = \rho C_p V_n d\theta(t), \tag{7}$$

$$= c_2 d\theta(t), \tag{8}$$

where

$$c_2 = \rho C_p 2\pi a^2 p, \tag{9}$$

and ρ denotes the density, C_p is the specific heat, and $d\theta(t)$ denotes the temperature rise.

Next, the total heat loss rate is given by:

$$\dot{Q}_L(t) = \dot{Q}_a(t) + \dot{Q}_r(t) \tag{10}$$

where $\dot{Q}_a(t)$ and $\dot{Q}_r(t)$ denote the axial and radial heat loss rates, respectively. Inserting their mathematical expressions, details of which can be found in [24–26], gives:

$$\begin{aligned} \dot{Q}_L(t) &= k_1\pi a^2\left[\frac{\theta(t)-\theta_I}{l_1} + \frac{10\theta(t)\beta L}{b\sqrt{\alpha}}\right] \\ &= \left[\frac{k_1\pi a^2}{l_1} + \frac{10k_1\pi a^2\beta L}{b\sqrt{\alpha}}\right]\theta(t) - \frac{k_1\pi a^2\theta_I}{l_1} \\ &= c_3\theta(t) - c_4 \end{aligned} \tag{11}$$

where

$$c_3 = \left[\frac{k_1\pi a^2}{l_1} + \frac{10k_1\pi a^2\beta L}{b\sqrt{\alpha}}\right], \tag{12}$$

$$c_4 = \frac{k_1\pi a^2\theta_I}{l_1}. \tag{13}$$

In the above equations, k_1 denotes the thermal conductivity, a is the nugget radius, and $\theta(t)$ and θ_I represent the melting temperature and the interface temperature of the workpieces, respectively.

Also, l_1 is the distance from the melting interface to the electrodes contact area, β represents the final penetration to workpiece thickness ratio, L is the sheet thickness, b represents the electrode radius, and α denotes the thermal diffusivity of the workpiece. Also, to avoid complexity of the model, the thermal conductivity and the thermal diffusivity are assumed to be constants.

The heat balance equation over the time interval, $(t, t + dt)$, is given by:

$$\dot{Q}_g(t) = \frac{H_f}{\Delta t}dt + dQ_T(t) + \dot{Q}_L(t)dt. \tag{14}$$

Substituting (1), (6), (8), and (11) in (14) and rearranging it, gives:

$$c_2\frac{d\theta(t)}{dt} = \frac{V^2(t)}{R(t)} - c_3\theta(t) + c_4 - c_1 \tag{15}$$

or equivalently,

$$\frac{d\theta(t)}{dt} = c_5\frac{V^2(t)}{R(t)} - c_6\theta(t) + c_7 \tag{16}$$

where

$$c_5 = \frac{1}{c_2}, \tag{17}$$

$$c_6 = \frac{c_3}{c_2}, \tag{18}$$

$$c_7 = (c_4 - c_1)/c_2 \tag{19}$$

Equation (16) represents a simplified electro-thermal dynamical model of the RSW process. From this, a dynamical resistance model is developed as follows.

3. Dynamical Resistance Model of an RSW Nugget Formation Process

The development of a dynamical resistance model exploits the functional relationship between resistance and temperature. To start with, it is assumed that $R(t)$ can be approximately represented by:

$$R(t) = R_\circ[1 + \alpha(\theta)(\theta(t) - \theta_\circ)], \tag{20}$$

where $\alpha(\theta)$ denotes the temperature coefficient of resistance, and $R_\circ$ is the resistance at room temperature, $\theta_\circ$.

Equivalently, one can write:

$$\theta(t) = \beta(\theta)(R(t) - R_\circ) + \theta_\circ \tag{21}$$

where

$$\beta(\theta) = \frac{1}{\alpha(\theta)R_\circ} \tag{22}$$

Differentiation of (21) yields:

$$\frac{d\theta(t)}{dt} = \beta(\theta)\frac{dR(t)}{dt} + \gamma(\theta)(R(t) - R_\circ) \tag{23}$$

where

$$\gamma(\theta) = \frac{d\beta(\theta)}{dt}. \tag{24}$$

Substitution of (16) in (23) gives:

$$\beta(\theta)\frac{dR(t)}{dt} + \gamma(\theta)(R(t) - R_\circ) = c_5\frac{V^2(t)}{R(t)} - c_6[\beta(\theta)(R(t) - R_\circ) + \theta_\circ] + c_7. \tag{25}$$

Also, re-arrangement of the above equation yields:

$$\frac{dR(t)}{dt} = c_8 \frac{V^2(t)}{R(t)} - c_9 R(t) + c_{10}(t), \tag{26}$$

where

$$c_8(t) = \frac{c_5}{\beta(\theta)}, \tag{27}$$

$$c_9(t) = \frac{\gamma(\theta)}{\beta(\theta)} + c_6, \tag{28}$$

$$c_{10}(t) = \frac{\gamma(\theta)}{\beta(\theta)} R_\circ + \frac{c_7}{\beta(\theta)} - \frac{c_6 \theta_\circ}{\beta(\theta)} + c_6 R_\circ \tag{29}$$

In Equations (27)–(29), the parameters c_8, c_9 and c_{10} are assumed to vary with time since $\beta(\theta)$ and $\gamma(\theta)$ vary with time (as a result of variation of θ with time).

Finally, for the sake of notational convenience, let $y(t) = R(t)$ and $u(t) = V(t)$. Then (26) can be rewritten as:

$$\frac{dy(t)}{dt} = c_8(t) \frac{u^2(t)}{y(t)} - c_9(t) y(t) + c_{10}(t) \tag{30}$$

Using a first order Euler approximation for $\frac{dy}{dt}$ with a sampling period, T_s, Equation (30) yields the following discrete time system:

$$\frac{y(k+1) - y(k)}{T_s} = c_8(k) \frac{u^2(k)}{y(k)} - c_9(k) y(k) + c_{10}(k) \tag{31}$$

or equivalently,

$$y(k+1) = A(k) y(k) + B(k) \frac{u^2(k)}{y(k)} + C(k) \tag{32}$$

where

$$A(k) = 1 - c_9(k) T_s, \tag{33}$$

$$B(k) = c_8(k) T_s, \tag{34}$$

$$C(k) = c_{10}(k) T_s, \tag{35}$$

and k is the discrete time index ($k = 0,\ 1,\ 2,\ \ldots$), with kT_s denoting the sampling instances.

Equation (32) represents a dynamical resistance model of the RSW process, which is characterized by three unknown time-varying parameters, *A*(*k*), *B*(*k*), and *C*(*k*). The validation of this model is discussed in the next subsection.

Validation of the Dynamic Resistance Model

The validation of the above dynamic resistance model requires voltage and current data from an adaptive weld controller that uses feedback signals to adjust its welding voltage with time. A set of voltage–current data collected from a constant heat controller (CHC) [26], manufactured by Welding Technology Corporation, is used for this purpose.

Figure 3 shows the welding voltage samples, V(k), and welding current samples, I(k), collected during a spot weld performed by a CHC machine. The dynamic resistance, R(k), is calculated from the above data by simply dividing V(k) by I(k) at each sample time, k. Then, a recursive least squares parameter estimation algorithm, described in Section 4.1 below, is used to estimate the parameters, A(k), B(k), and C(k), of the above dynamic resistance model. These estimated model parameters are shown in Figure 4.

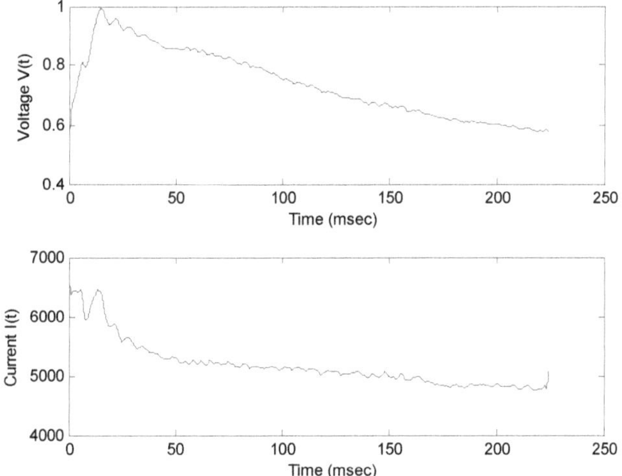

Figure 3. Voltage, V(t), and current, I(t), from a weld performed by a CHC controller.

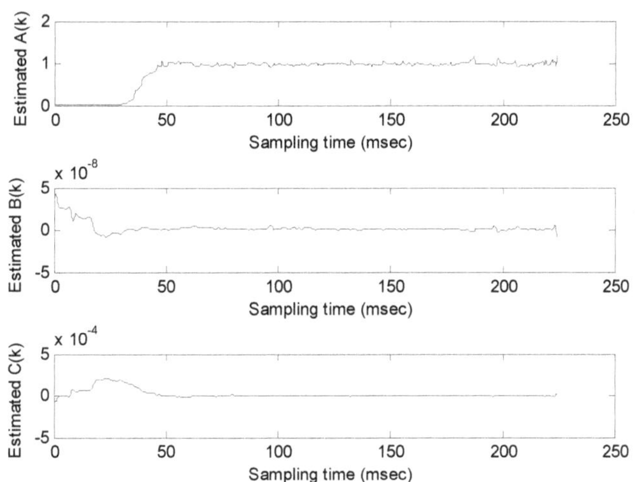

Figure 4. Estimated model parameters from the voltage and current data shown in Figure 3.

Finally, in order to assess the goodness of fit of the proposed model, the estimated model parameters, $\hat{A}(k)$, $\hat{B}(k)$, and $\hat{C}(k)$, are next used to compute the predicted resistance values, $R_p(k)$, at each sample point, k. The goodness of fit between actual resistance, R(k), and the model-predicted resistance, $R_p(k)$, is shown in Figure 5, which demonstrates the validity of the above model. It may be pointed out that similar results were also obtained from other CHC welding data.

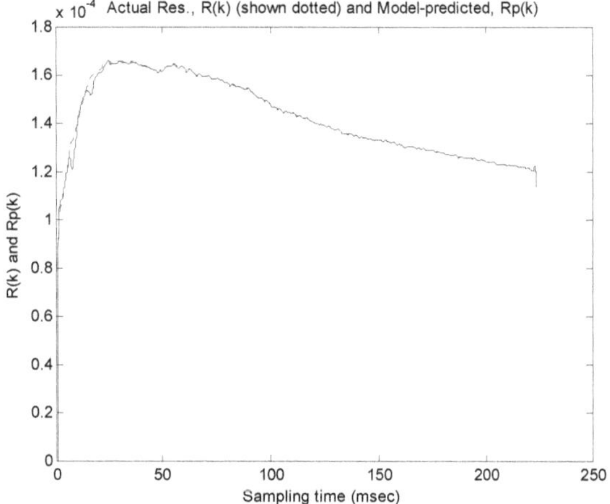

Figure 5. Goodness of fit between actual dynamical resistance (shown dotted), R(t), and its model-predicted values, $R_p(t)$.

4. Design of an Adaptive RSW Controller

Since the system represented by Equation (32) is characterized by time-varying parameters, an adaptive controller would be an appropriate tool for controlling such a system. However, the nonlinear nature of the above system precludes usage of a conventional linear adaptive controller. In view of above, an adaptive one-step-ahead controller is proposed to be used here. The three key steps required to implement such a controller involve measurement of the input (welding voltage) and output (dynamic resistance) at uniform sampling intervals, estimation of the dynamic resistance model parameters, $\hat{A}(k)$, $\hat{B}(k)$, and $\hat{C}(k)$, using a recursive parameter estimation algorithm, such as recursive least squares (RLS), and computation of a control signal based on the estimated parameter values. The estimation of model parameters and computation of a control signal are discussed in the following subsections.

4.1. Parameter Estimation

First, the model Equation (32) is rewritten in the following predictive form:

$$y(k+1) = \varphi(k)^T X^* \tag{36}$$

where

$$\varphi(k) = \begin{bmatrix} y(k) & \frac{u^2(k)}{y(k)} & 1 \end{bmatrix}^T, \tag{37}$$

$$X^* = \begin{bmatrix} A & B & C \end{bmatrix}^T. \tag{38}$$

Here, X denotes the vector of unknown system parameters that are estimated recursively using the following RLS algorithm:

$$\hat{X}(k) = \hat{X}(k-1) + \frac{P(k-2)\varphi(k-1)}{\lambda + \varphi(k-1)^T P(k-2)\varphi(k-1)}\left[y(k) - \varphi(k-1)^T \hat{X}(k-1)\right];\ k \geq 1, \tag{39}$$

$$P(k-1) = \left(\frac{1}{\lambda}\right)[P(k-2) - \frac{P(k-2)\varphi(k-1)\varphi(k-1)^T P(k-2)}{1 + \varphi(k-1)^T P(k-2)\varphi(k-1)}], \tag{40}$$

$$\hat{X}(0) = \begin{bmatrix} 0 & \varepsilon & 0 \end{bmatrix}^T, \tag{41}$$

$$P(-1) = \sigma I, \tag{42}$$

where λ, $0 < \lambda < 1$, denotes a forgetting factor, $\varepsilon > 0$ is a small number, and $\hat{B}(k)$ is always constrained to be non-negative, i.e.,

$$\hat{B}(k) \geq \varepsilon > 0 \text{ for all } k. \tag{43}$$

Furthermore, since $\hat{X}(k)$ denotes an estimate of X^*, the predicted output at time $k+1$ is defined as:

$$\hat{y}(k+1) = \varphi(k)^T \hat{X}(k). \tag{44}$$

4.2. Adaptive One-Step-Ahead Tracking Controller

A one-step-ahead (OSA) RSW control scheme based on a thermal model of the RSW process was investigated in [23,24], and a similar methodology is followed here. An OSA controller attempts to bring the predicted output, $y(k+1)$ at time $k+1$, to the desired value, $y^*(k+1)$, in one step. Thus, it minimizes the following cost function [27]:

$$J_1(k+1) = \frac{1}{2}[y(k+1) - y^*(k+1)]^2 \tag{45}$$

The corresponding OSA control law is given by:

$$\overline{u}^2(k) = \frac{y^*(k+1)y(k) - A(k)y^2(k) - C(k)y(k)}{B(k)} \tag{46}$$

or equivalently,

$$\overline{u}(k) = \sqrt{\frac{y^*(k+1)y(k) - A(k)y^2(k) - C(k)y(k)}{B(k)}}. \tag{47}$$

The above control signal needs to be constrained by the maximum voltage delivery capacity of the weld controller, u_{max}, as follows:

$$u(k) = \begin{cases} \overline{u}(k), & if\ 0 < \overline{u}(k) < u_{max} \\ 0, & if\ \overline{u}(k) \leq 0 \\ u_{max}, & if\ \overline{u}(k) \geq u_{max} \end{cases} \tag{48}$$

The adaptive OSA controller uses the estimate, $\hat{X}(k)$, in Equation (39) to compute the control signal, $\overline{u}(k)$, from the following adaptive version of Equation (47):

$$\overline{u}(k) = \sqrt{\frac{y^*(k+1)y(k) - \hat{A}(k)y^2(k) - \hat{C}(k)y(k)}{\hat{B}(k)}} \tag{49}$$

where $\hat{A}(k)$, $\hat{B}(k)$, and $\hat{C}(k)$ denote the estimated values of $A(k)$, $B(k)$, and $C(k)$, respectively, at time k.

Remark

Since welding times are usually very short (less than 0.5 sec), establishing proof of asymptotic tracking would be meaningless here. However, since the input voltage and output resistance are bounded in this case, the regression vector, $\varphi(k)$, in Equation (36) is bounded for all k. This ensures that the RLS parameter estimation algorithm, described by Equations (39)–(42), possesses nice convergence properties [27] that help the output resistance track the reference resistance profile. Furthermore, since most RSW applications require successive welds under very similar conditions, tracking can be significantly improved by initializing the RLS parameter estimator to the nominal values of the process parameters, which essentially remain constant or change very slowly from one weld to the next.

5. Simulation Results and Discussion

The results of a simulation study to evaluate the performance of the proposed controller and compare it to that of a PID controller are presented in this section. Both controllers are designed for tracking a desired reference resistance profile.

As mentioned earlier, a reference resistance profile serves as a good indicator of the weld quality. Therefore, it is desirable to maintain the dynamic resistance of a (forming) weld nugget reasonably close to a reference resistance profile. For these simulations, two sheets of 1.2-mm-thick mild steel were used as the materials to be welded. The properties of this material and the RSW model parameters are listed in Table 1 below.

Table 1. Material properties and RSW model parameters.

Symbol	Description	Value	Units
ρ	Density	7800	Kg.m^{-3}
C_p	Specific Heat	480	J.Kg^{-1}.C^{-1}
a	Nominal Nugget Radius	2.50×10^{-3}	m
p	Nominal Nugget Penetration per Sheet	1.08×10^{-3}	m
k	Thermal Conductivity	15.1	W.m^{-1}.C^{-1}
l_1	Nominal Indentation per Sheet	1.20×10^{-4}	m
β	Penetration to Workpiece Thickness Ratio	0.8	no units
L	Sheet Thickness	1.20×10^{-3}	m
b	Electrode Radius	3.00×10^{-3}	m
L_f	Latent Heat of Fusion	2.73×10^{5}	J.Kg^{-1}
H	Heat of Fusion	2.13×10^{9}	J.m^{-3}
α	Thermal Diffusivity	4.03×10^{-6}	m^2.s^{-1}
θ_I	Interface Temperature	500	°C
$\theta_\circ$	Initial Temperature	20	°C
α_r	Temperature Coefficient	0.0066	°C^{-1}
$\rho_\circ$	Resistivity @ 20 °C	0.15	µΩ.m

A typical reference resistance profile for a good weld made on such a workpiece is shown in Figure 6, and it is used as a reference in this study. Depending on the error signal, the welding voltage is adjusted so as to reduce the resistance tracking error. Also, the controller is assumed to be capable of delivering a maximum voltage, $V_{max} = 1600$ mV.

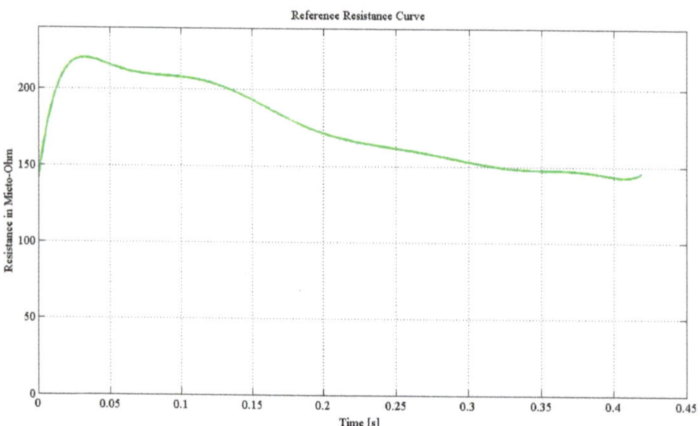

Figure 6. Desired reference resistance profile.

Although RSW welds are usually performed under constant force, some variations of the applied pressure always occur during welding. Also, electrode tips exhibit wear and tear due to repeated hammering and deposition of metals on the tip surfaces. In view of above, force variation and electrode wear are considered to be unknown process (noise) variables that impact the nugget size (diameter and penetration). Since an RSW weld controller must be evaluated in the presence of such ubiquitous process (noise) variables, the performance of an AOSA controller and a PID controller are tested in case of a 20% increase in nugget diameter and a 50% increase in indentation from their nominal values.

Figure 7 shows the performance of the AOSA controller. It can be seen that the AOSA controller adapts to the parameter changes and enables the output resistance profile to follow the desired resistance profile reasonably well. In fact, tracking can be further improved by initializing the RLS parameter estimator to the nominal values of the process parameters. Also, the energy required to perform the weld is found to be 1180 J.

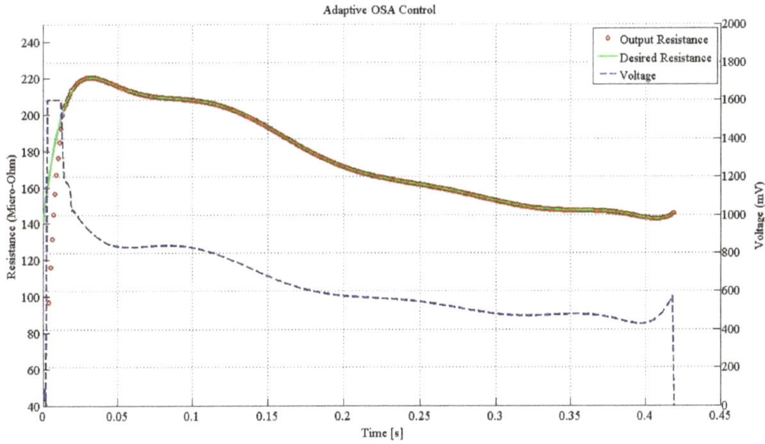

Figure 7. Tracking performance of the AOSA controller. The reference resistance profile is shown as a continuous (green) curve and the actual (output) resistance is shown in (orange) circles. The control (welding) voltage, V(t), is shown as a dashed (blue) curve ($V_{max} = 1600$ mV). Energy consumed to perform the weld = 1180 J.

Figure 8 shows the performance of a PID controller. After a number of trials to tune the performance of the PID controller, the optimal gains of the controller were chosen to be as follows: Proportional (P) = 0.023, Integral (I) = 3.14, Derivative (D) = −6.99. At its best performance, it can be seen that the PID controller initially loses track of the reference resistance profile due to variation of welding parameters, which can be detrimental to nugget formation. Also, it can be seen that the PID controller requires more energy (1217 J) to perform the weld, as compared to AOSA.

Comparing the simulation results for the above controllers, the AOSA controller is seen to exhibit satisfactory performance and, in this case, the output resistance profile tracks the desired reference resistance profile quite well. Also, it may be noted that the total energy required to perform a weld using an AOSA controller is less compared to that used by a PID controller. In the long run, this can yield significant energy savings for applications requiring a high volume of spot welds, such as manufacturing of automotive vehicles.

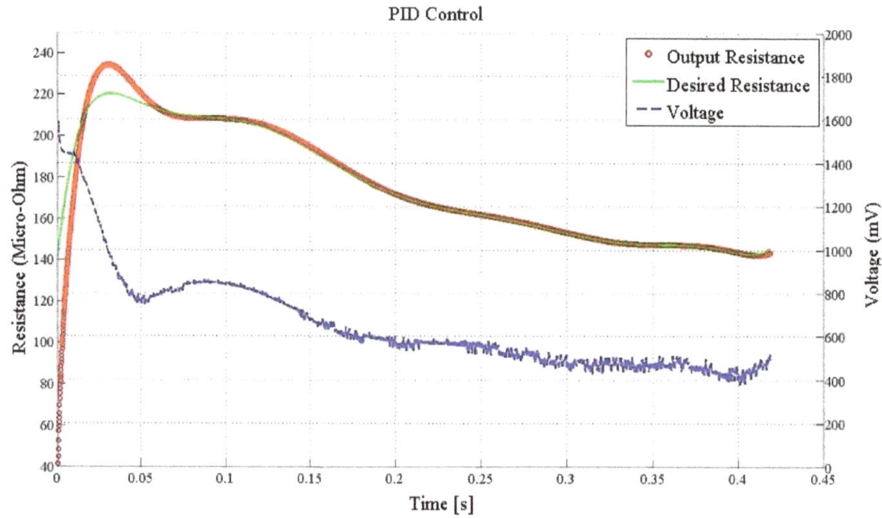

Figure 8. Tracking performance of the (tuned) PID controller. The reference resistance profile is shown as a continuous (green) curve and the actual (output) resistance is shown in (red) circles. The control (welding) voltage, V(t), is shown as a dashed (blue) curve ($V_{max} = 1600$ mV). Energy consumed to perform the weld = 1217 J.

6. Conclusions

This paper proposes a new theoretical approach of designing an AOSA controller for resistance spot welding that utilizes a simplified dynamic resistance model of an RSW process. The development of this model and its validation are discussed in detail. It also presents the results of a simulation study that compares the performance of the proposed AOSA controller with that of a PID controller. The simulation results indicate that an AOSA controller is capable of compensating for some process parameter variations, and also enables tracking of a desired reference resistance profile. Also, these results indicate that an AOSA controller is capable of reducing the energy consumed per weld, which may yield significant energy and cost savings for applications requiring a high volume of spot welds. It should be mentioned, however, that this paper only lays the foundation of a viable adaptive RSW control scheme based on a dynamic resistance model. Actual hardware implementation and testing of the proposed scheme is currently under investigation and results of these studies are expected to be presented in a follow-up paper.

Author Contributions: Conceptualization and methodology, Z.K. and M.D.; investigation, Z.K. and M.D.; writing and editing, Z.K. and M.D.

Funding: This research received no external funding.

Conflicts of Interest: The authors declare no conflicts of interest.

References

1. Zhang, H.; Senkara, J. *Resistance Welding Fundamentals and Applications*; Taylor & Francis Group: Boca Raton, FL, USA, 2012.
2. Govik, A. Modeling of the Resistance Spot Welding Process. Master Thesis, Linköping University, Linköping, Sweden, 2009.
3. Won, Y.J.; Cho, H.S.; Lee, C.W. A Microprocessor-Based Control System for Resistance Spot Welding Process. In Proceedings of the American Control Conference, San Francisco, CA, USA, 22–24 June 1983; pp. 734–738.
4. Zhou, K.; Cai, L. A Nonlinear Current Control Method for Resistance Spot Welding. *Proc. ASME Trans. Mechatron.* **2014**, *19*, 559–569. [CrossRef]

5. Salem, M.; Brown, L.J. Improved Consistency of Resistance Spot Welding with Tip Voltage Control. In Proceedings of the 24th Canadian Conference on Electrical and Computer Engineering, Niagara Falls, ON, Canada, 8–11 May 2011; pp. 548–551.
6. Chen, X.; Araki, K.; Mizuno, T. Modeling and Fuzzy Control of the Resistance Spot Welding Process. In Proceedings of the 24th Canadian Conference on Electrical and Computer Engineering, Tokushima, Japan, 29–31 July 1997; pp. 898–994.
7. El-Banna, M.; Filev, D.; Chinnam, R.B. Intelligent Constant Current Control for Resistance Spot Welding. In Proceedings of the IEEE Conference on Fuzzy Systems, Vancouver, BC, Canada, 16–21 July 2006; pp. 1570–1577.
8. Chen, X.; Araki, K. Fuzzy Adaptive Process Control of Resistance Spot Welding with a Current Reference Model. In Proceedings of the IEEE Conference on Intelligent Processing Systems, Beijing, China, 28–31 October 1997; pp. 190–194.
9. Shriver, J.; Peng, H.; Hu, S.J. Control of Resistance Spot Welding. In Proceedings of the 1999 American Control Conference, San Diego, CA, USA, 2–4 June 1999; pp. 187–191.
10. Ivezic, N.; Allen, J.D.; Zacharia, T. Neural Network-Based Resistance Spot Welding Control and Quality Prediction. In Proceedings of the Second International Conference on Intelligent Processing and Manufacturing of Materials, Honolulu, HI, USA, 10–15 July 1999; pp. 989–994.
11. Messler, R.W.; Jou, M.; Li, C.J. An Intelligent Control System for Resistance Spot Welding Using a Neural Network and Fuzzy Logic. In Proceedings of the 1995 IEEE Industry Applications Conference Thirtieth IAS Annual Meeting, Orlando, FL, USA, 8–12 October 1995; pp. 1757–1763.
12. Yu, J. New methods of resistance spot welding using reference waveforms of welding power. *Int. J. Precis. Eng. Manuf.* **2016**, *17*, 1313–1321. [CrossRef]
13. Subbanna, R.S.; Rajini, M. Intelligent Control System for Resistance Spot Welding System. *Manag. J. Instrum. Control Eng.* **2018**, *6*, 35–41. [CrossRef]
14. Yu, J. Adaptive Resistance Spot Welding Process that Reduces the Shunting Effect for Automotive High-Strength Steels. *Metals* **2018**, *8*, 775. [CrossRef]
15. Dickinson, D.; Franklin, J.; Stanya, A. Characterization of Spot-Welding Behavior by Dynamic Electrical Parameter Monitoring. *Weld. J.* **1980**, *59*, S170–S176.
16. Kaiser, J.; Dunn, D.; Eagar, T. The Effect of Electrical-Resistance on Nugget Formation during Spot-Welding. *Weld. J.* **1982**, *61*, S167–S174.
17. Garza, F.; Das, M. On real time monitoring and control of resistance spot welds using dynamic resistance signatures. In Proceedings of the 44th IEEE 2001 Midwest Symposium on Circuits and Systems, Dayton, OH, USA, 14–17 August 2001.
18. Cho, Y.; Rhee, S. Primary Circuit Dynamic Resistance Monitoring and its Application to Quality Estimation during Resistance Spot Welding. *Weld. J.* **2002**, *81*, S104–S111.
19. Wan, X.; Wang, Y.; Zhao, D. Quality monitoring based on dynamic resistance and principal component analysis in small scale resistance spot welding process. *Int. J. Adv. Manuf. Technol.* **2016**, *86*, 3443–3451. [CrossRef]
20. Luo, Y.; Rui, W.; Xie, X.; Zhu, Y. Study on the nugget growth in single-phase AC resistance spot welding based on the calculation of dynamic resistance. *J. Mater. Process. Technol.* **2016**, *229*, 492–500. [CrossRef]
21. Podrzaj, P.; Polajnar, I.; Diaci, J.; Kariz, Z. Overview of resistance spot welding control. *Sci. Technol. Weld. Join.* **2008**, *13*, 215–224. [CrossRef]
22. Hemmati, M.; Haeri, M. Control of resistance spot welding using model predictive control. In Proceedings of the 2015 9th International Conference on Electrical and Electronics Engineering, Bursa, Turkey, 26–28 November 2015; pp. 864–868.
23. Kas, Z.; Das, M. A Thermal Dynamical Model Based Control of Resistance Spot Welding. In Proceedings of the IEEE International Conference on Electro/Information Technology, Milwaukee, WI, USA, 5–7 June 2014; pp. 264–269.
24. Kas, Z.; Das, M. An Electrothermal Model Based Adaptive Control of Resistance Spot Welding Process. *Intell. Control Autom.* **2015**, *6*, 134–146. [CrossRef]

25. Kim, E.W.; Eagar, T.W. Parametric Analysis of Resistance Spot Welding Lobe Curve. *SAE Trans. J. Mater.* **1988**. [CrossRef]
26. CHC Algorithm, WTC. Available online: http://www.weldtechcorp.com/welding_concepts/weldsolutions_chc.html (accessed on 27 September 2019).
27. Goodwin, G.C.; Sin, K.S. *Adaptive Filtering Prediction and Control*; Prentice-Hall: Englewood Cliffs, NJ, USA, 1983.

© 2019 by the authors. Licensee MDPI, Basel, Switzerland. This article is an open access article distributed under the terms and conditions of the Creative Commons Attribution (CC BY) license (http://creativecommons.org/licenses/by/4.0/).

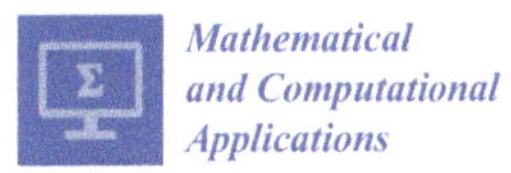

Article

Hybrid State Constraint Adaptive Disturbance Rejection Controller for a Mobile Worm Bio-Inspired Robot

Vania Lara-Ortiz [1], Ivan Salgado [1], David Cruz-Ortiz [2,3], Alejandro Guarneros [1], Misael Magos-Sanchez [1] and Isaac Chairez [3,*]

[1] CIDETEC, Instituto Politécnico Nacional, 07738 Ciudad de México, Mexico; vaniip19@gmail.com (V.L.-O.); ijesusr@gmail.com (I.S.); alejandrowarnsan@gmail.com (A.G.); je11dat6@hotmail.es (M.M.-S.)

[2] Departamento de Control Automático, CINVESTAV-IPN, 07360 Ciudad de México, Mexico; cuod.cruz.ortiz@gmail.com

[3] UPIBI, Instituto Politécnico Nacional, 07340 Ciudad de México, Mexico

* Correspondence: ichairezo@gmail.com or jchairezo@ipn.mx; Tel.: +52-5514-01-93074

Received: 31 December 2019; Accepted: 29 February 2020; Published: 4 March 2020

Abstract: This study presents the design of a hybrid active disturbance rejection controller (H-ADRC) which regulates the gait cycle of a worm bio-inspired robotic device (WBRD). The WBRD is designed as a full actuated six rigid link robotic manipulator. The controller considers the state restrictions in the device articulations; this means the maximum and minimum angular ranges, to avoid any possible damage to the structure. The controller uses an active compensation method to estimate the unknown dynamics of the WBRD by means of an extended state observer. The sequence of movements for the gait cycle of a WBRD is represented as a class of hybrid system by alternative reference frameworks placed at the first and the last link. The stability analysis employs a class of Hybrid Barrier Lyapunov Function to ensure the fulfillment of the angular restrictions in the robotic device. The proposed controller is evaluated using a numerical simulation system based on the virtual version of the WBRD. Moreover, experimental results confirmed that the H-ADRC may endorse the realization of the proposed gait cycle despite the presence of perturbations and modeling uncertainties. The H-ADRC is compared against a proportional derivative (PD) controller and a proportional-integral-derivative (PID) controller. The H-ADRC shows a superior performance as a consequence of the estimation provided by the homogeneous extended state observer.

Keywords: adaptive disturbance rejection controller; hybrid systems; state constraint; worm robot; bio-inspired robots

1. Introduction

The main objective of bio-mimetics is to find a practical solution of human needs imitating models or movements of animals or even plants. One of its main applications can be found in the field of robotics [1]. The development of bio-inspired robotic systems involves the adaptation of different modes of locomotion like the running inspired in leopards [2], swimming inspired in fishes [3], climbing like gecko robots [4], or crawling by worms [5,6], among others. In the case of worm bio-inspired robots, the movement of the so-called *inchworm* has interesting applications exploring narrow places in contrast to mobile robots [7]. The inchworm moves with a looping movement in which the anterior and posterior legs are alternately made fast and released. The alternation of fastening enables a propelling motion [8]. These bio-inspired robots can be applied in medical applications like colonoscopies [9], in the inspection of narrow pipes [7] and robotic manipulators [10]. There exist diverse configurations of inchworm robots from two DOF (Degrees of freedom) to five. Two DOF in-pipe robots such as [7] reproduces contraction and expansion of inchworm's gait cycle using two sets of magnetic clamps switching an electro-valve: rear clamp grasps the pipe firmly while the front clamp slides forward gaining traction in the process. Similarly, in [11], a system of two mass with a spring that contracts/expands by its anisotropic skin is described. The inching mechanism was proposed also in [12] for planetary surface exploration vehicles (rovers) to overcome the limitations of traditional rolling mobility. The vehicle wheel bases were expanded and contracted to achieve an increase of net traction potential. In addition, in [13], a three-module bore robot was constructed to carry out investigations on planetary subsurfaces such as geothermal gradient, chemical composition and analysis of regolith. *Climbot* is a tele-operated five DOF (Degrees) robot able to climb a variety of media and grasp objects [14].

One interesting problem to solve is the tracking trajectory problem in these kinds of robots [15]. The complex structures that emulate the displacement of an inchworm bio-inspired robot require robust techniques to cope with parametric uncertainties, no modeled dynamics, and noisy measurements. Classical PID controllers, sliding modes, and fuzzy logic controllers have been applied without considering the hybrid behavior of the gait cycle of an inchworm represented by a multi-link robot manipulator [16]. Modeling an inchworm robot that alternates the grasping between its anterior and posterior legs implies a switching structure that should be studied under the concept of hybrid systems.

The hybrid framework allows for studying more complex dynamics and allows more flexibility in modeling dynamic phenomena [17]. A hybrid system is composed of two or more sets of differential equations describing a particular stage of behavior in dynamic systems. In the case of the WBRD, two robot manipulators with five DOF represent its gait cycle. In order to deal with non-modeled dynamics and parameter uncertainties, an active disturbance rejection (H-ADRC) approach can be considered. ADRC is a technique centered on providing an effective estimation of unknown nonlinearities by means of algebraic techniques [18]. The main concepts considered in this control designs are: (1) simplify the plant description so as to group all disturbances and uncertainties, as well as all unknown or ignored quantities and expressions into a single disturbance term, (2) proceed to estimate the effects of this disturbance, in some accurate manner, and (3) devise the means to cancel its effects, using a feasible gathered estimate as part of the feedback control action. One way to fulfill this task is to perform a polynomial expansion and translate it into the state space as the output of an extended state observer [19].

The control algorithm has to be able not only to force the WBRD to reach a desired trajectory; it needs to take into account the problem of finite-time convergence and states constraints to avoid any damage of the mechanical structure. A classical tool to deal with state constraints is the concept of Barrier Lyapunov functions (BLF) that is a function that tends to infinite as the argument approach to a boundary. BLF has been applied to control nonlinear systems and linear perturbed systems [20].

This manuscript proposes a novel adaptive algorithm to deal with the trajectory tracking problem of nonlinear hybrid systems with state constraints. The proposed algorithm is applied in the WBRD with five DOF represented by a hybrid structure. The main contributions of this study are:

- The mechanical design of a bio-inspired inchworm robot with a hybrid structure.
- A hybrid ADRC controller capable of estimating the non-modeled dynamics and providing the fundamentals to prove the origin of the tracking error space is a practical stable equilibrium point considering the effect of the presence of non-modeled dynamics and state constraints.
- The complete stability analysis with a BLF providing ultimate boundedness for the tracking error.
- An additional complementary adaptive algorithm to reduce the energy consumed by the controller.
- The experimental confirmation of the controller application on an instrumented WBRD that may emulate a gait cycle of a particular inchworm.

This manuscript is organized in the following manner. Section 2 provides a general overview of the WBRD design as well as the links–joints configuration. Section 3 introduces the control design problem statement considering the hybrid nature of the gait cycle realization by the WBRD. Section 4 provides the formulation of the WBRD realization in terms of the hybrid systems framework. The next Section 5 details all the elements of the output feedback controller to solve the gait cycle of the WBRD. Section 6 describes some aspects regarding the implementation (numerical and experimental) of the output feedback controller. Section 7 provides the evidence of the controller numerical implementation over a virtualized representation of the WBRD. Section 8 demonstrates the application of the suggested controller on a developed WBRD using the tri-dimensional (3D) printing technique. Finally, Section 9 closes the paper with some final remarks.

2. Worm Bio-inspired Robotic Device

The proposed WBRD structure satisfies a class of multi-articulated manipulator with 5 DOF. The WBRD displacement is realized by the switched fixation of the non-inertial frames (1st and 5th) to the supporting surface (Figure 1). Considering that WBRD moves following a path tracking based on sequenced steps, the odd steps occur with the 1st frame as the reference and the even steps happened considering the 5th frame as reference. The sequence formed by odd-even steps defines a gait cycle of the WBRD. As one may notice, the change of the reference frame justifies the use of switched systems theory to develop the output feedback controller to regulate the WBRD mobilization. This can be noticed with the alternated reference frame marked with black squares at the bottom of Figure 1.

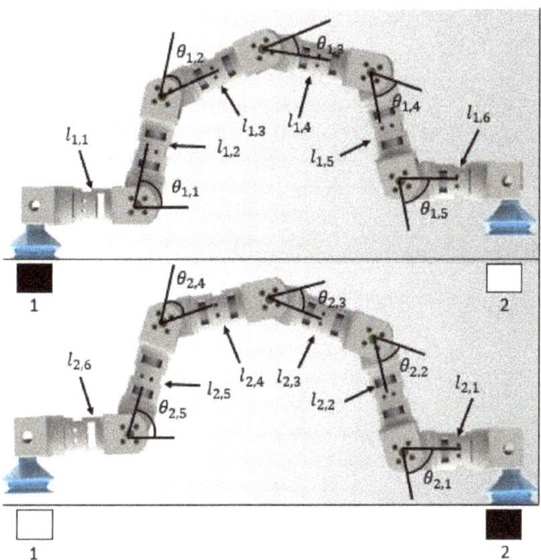

Figure 1. Distribution of the WBRD's degrees of freedom dependent on the activated vacuum pump.

The multi-articulated manipulator is formed by five solid links $l_{s,i}$ connected with rotational joints characterized with angular displacement defined by $\theta_{s(t),i}$. The variable s can be 1 if the step is odd and 2 if the step is even. This variable is playing the role of the switching sequence usually considered in switching systems analysis. The switching action is performed by a set of vacuum pumps that emulates the front and rear legs subjection to the floor (see Figure 2). Each link is conformed by a direct current (DC) motor for actuation, and a set of mechanical elements to transmit the movements. To obtain a feedback of the actual position of each link, a set of five markers were placed in the robot. A vision analysis system obtained the corresponding absolute angles of each link. These measurements were the data input into the control algorithm.

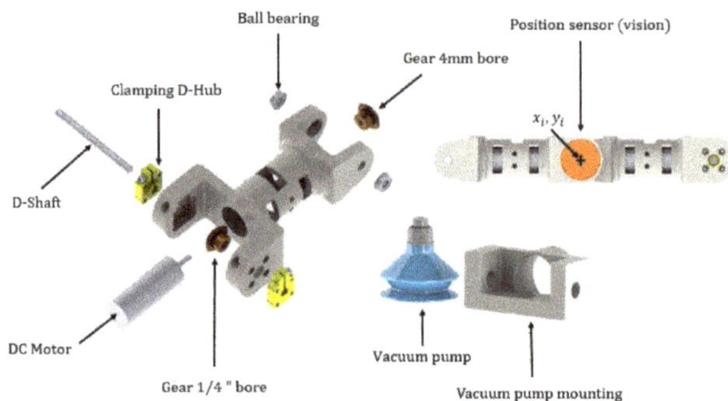

Figure 2. Elements of each link in the WBRD.

Table 1 describes the dimensions of the angles.

Table 1. Angular range of the WBRD.

Element		Min	Max	Range
$\theta_{1,1}$	$\theta_{2,5}$	−80	80	160
$\theta_{1,2}$	$\theta_{2,4}$	−95	95	190
$\theta_{1,3}$	$\theta_{2,3}$	−95	95	190
$\theta_{1,4}$	$\theta_{2,2}$	−95	95	190
$\theta_{1,5}$	$\theta_{2,1}$	−80	80	160

Each link was designed with a scale of 50 : 1 yielding a total length of 71.6 cm in the zero position and a total height of 7.2 cm including the pumps (Figure 3).

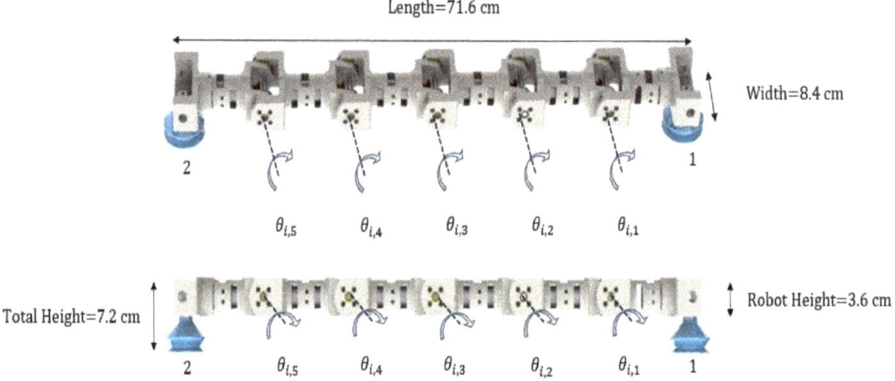

Figure 3. Dimensions of the WBRD.

3. Problem Definition

The WBRD displacement is represented as an alternated extension–contraction sequences (Figure 4). This simplified representation of the WBRD displacement can be described as a hybrid device alternating the movement of two multi-articulated (5 DOF) manipulators.

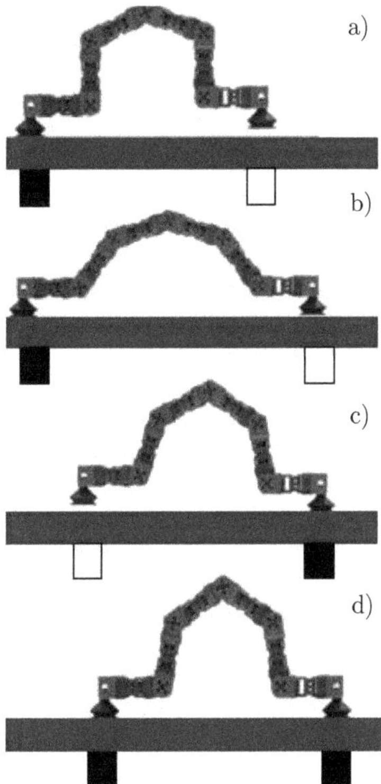

Figure 4. Alternative representation of a gait cycle for the WBRD; (**a**) The rear pump is *on* and a first robot manipulator structure is adopted; (**b**) The reference trajectories force the robot to expand almost to reach a 0 degrees configuration; (**c**) The pumps switch and the second robot manipulator configuration is adopted, the reference trajectories for the movement of the robot until a desired position; (**d**) A second switching in the gait cycle is performed to complete the walking path emulating the real inchworm.

The fixation of the reference frame in the suggested alternate way modifies the description of the WBRD. Indeed, this variation of the reference frame forces two distinct dynamic representations for the WBRD. Such condition provides a challenging scenario for developing automatic controllers which can ensure the tracking of reference trajectories that correspond to a bio-inspired gait cycle. This section aims to formulate the controller design problem within the hybrid systems' framework.

Let us consider the vector of angular displacements within a fixed part ($s(t)$ equal either a or b of the gait cycle $\theta_{s(t)} = |\theta_{s(t),i}|_{i=1,\ldots,5}$. Now, assume that, during the given part of the gait cycle, the angular displacements must track the corresponding reference angles $\theta^*_{s(t)} = |\theta^*_{s(t),i}|_{i=1,\ldots,5}$. Then, enforcing the gait cycle for the WBRD can be represented as an stabilization problem for the tracking error $\Delta \in \mathbb{R}^5$, defined as $\Delta_{s(t)} = \theta_{s(t)} - \theta^*_{s(t)}$ within each continuous domain of $s(t)$.

This problem statement obligates to consider the WBRD dynamics changes only if the vector of tracking errors for all the articulations has attained a sufficiently small value (defined by the user), namely $SW^* > 0$. Therefore, the triggering signal which enforces the dynamics changing can be obtained by measuring the norm of the tracking error within each domain of $s(t)$. Once $\|\Delta_{s(t)}\| \leq SW^*$, then $s(t)$ changes from a to b or vice versa.

The problem statement concept given above enforces the fact that the vector of the angular displacements $\theta_{s,i}$ at the WBRD must track the desired reference angles $\theta^*_{s,i}$ ensuring the tracking errors of all articulations enter the region characterized by $\|\Delta_{s(t)}\| \leq SW^*$ at some given finite moment T_c

which must be bounded ($T_s > 0$). Such tracking problem can be described as designing the hybrid controller $u_{s(t)}$ such that

$$\|\Delta_{s(t)}(t)\| \leq SW^* \qquad \forall t \in [T_c, T_s] \tag{1}$$

The maximum allowed switching time is introduced here in order to have a tracking trajectory independent safety condition that can turn off the WBRD if the switching condition is not attained in a reasonable period. Notice that the switching condition introduces a class of a non-constant sampling discrete state which depends on the accomplishment of the condition provided in (1).

This study assumes that only $\theta_{s(t)}$ is continuously locally measurable all the time. However, $\dot{\theta}_{s(t)}$ is not available. Therefore, the control design considers an output feedback realization.

4. Hybrid Formulation of the Worm Walking Cycle

The changing dynamics of WBRD can be characterized using a combination of continuous and discrete states. Such representation agrees with the fundamentals of hybrid systems [21]. This formulation to describe the gait evolution of the proposed WBRD is enforced because there is not a strict periodicity which may define the transition between the gait domains (a or b) that is from continuous to continuous (a to b or vice versa) dynamics passing through the discrete state domain.

If the WBRD exerts a regular mono-directional walking gait, the transitions between the continuous stages follows an ordered sequence (a -> b -> a ->...). This sequenced dynamical behavior justifies the application of a class of multi-domain hybrid systems framework considering a predefined order of phases (or domains). Such representation leads to defining a so-called coherent cycle.

Formally, a multi-domain hybrid control system can be described considering a tuple [22,23] $\mathcal{HD} = (\Gamma, D, U, S, \Delta, FG)$, where $\Gamma = (V; E)$ describes the sequenced cycle of transitions. Consequently, $v \in V$ defines a transition vertex connection the continuous domains, v^+ represents the subsequent vertex of v during the gait cycle, while $e = \{v \to v^+\}$ corresponds to the transition from the analyzed vertex v to v^+.

Continuous Dynamical Representations. Considering the links masses, their inertia as well as their lengths properties of the WBRD, the equation of motion (EOM) that can be used within a given continuous domain D_v can be determined by the Euler–Lagrange equations (considering that, within a given domain, the WBRD obeys a manipulator representation) [24]. Therefore, assuming that $x_a = \theta_{s(t)}$ in each fixed domain, the dynamics of the WBRD corresponds to:

$$\begin{gathered}\dot{x}_a(t) = x_b(t) \\ \dot{x}_b(t) = f\left(x_a(t), x_b(t)\right) + g\left(x_a(t)\right)u(t) + D^{-1}\left(x_a(t)\right)J_v^{\top}\left(x_a(t)\right)F_v\left(x_a(t), x_b(t)\right) + \\ \xi\left(x_a(t), x_b(t), t\right)\end{gathered} \tag{2}$$

Here, $x_a \in \mathbf{X}_a \subset \mathbb{R}^5$, $x_b \in \mathbf{X}_b \subset \mathbb{R}^5$ and $u \in \mathbb{R}^n$ are the vectors of angular displacements, angular velocities, and the applied torques (operating as the control actuators) respectively for the WBRD.

The drift vector field $f : \mathbf{X}_a \times \mathbf{X}_b \to \mathbb{R}^5$ corresponds to $f(x_a, x_b) = -D^{-1}(x_a)\left[C(x_a, x_b)x_b + G(x_a)\right]$, the state dependent matrix $D : \mathbf{X}_a \to \mathbb{R}^{5\times5}$ defines the inertia of the WBRD, the matrix $C : \mathbb{R}^5 \times \mathbb{R}^5 \to \mathbb{R}^{5\times5}$ defines the Coriolis effects while $G : \mathbb{R}^5 \to \mathbb{R}^5$ defines the effect of gravitational force over the WBRD dynamics. The vector function $g : \mathbb{R}^5 \to \mathbb{R}^{5\times5}$ characterizes the control action effect over the WBRD dynamics with $g = D^{-1}$.

The uncertain section of the model is gathered in $\xi : \mathbf{X}_a \times \mathbb{R}^{n_v} \times \mathbb{R} \to \mathbb{R}^5$ which characterizes the presence of external perturbations and internal modeling uncertainties in the WBRD. Usually, this term aggregates nonlinear behavior such as joint frictions, backslash, and some other elements that are usually complex for modeling.

The function $F_v : TQ \times U_v \to \mathbb{R}^{n_v}$ (n_v is the number of total holonomic restrictions) represents the contact wrenches containing the constraint forces and/or moments. Here, TQ represents the characteristic states occurring during the contact wrenches, and U_v is the corresponding set of control actions which leads to the contact wrenches u_v, which are relevant during the WBRD transition from continuous to continuous domains (floor contact). To enforce the velocity independent (holonomic) constraints, the second order differentiation of the constraints should be set to zero; that is,

$$J_v(x_a)\dot{x}_b + \dot{J}_v\left(x_a, x_b\right)x_b = 0 \tag{3}$$

The constrained dynamics of the system must be determined using the trajectories of (2) together with (3).

Holonomic Constraints. Given that the WBRD model with coordinates $x_a \in Q$, $Q \in \mathbb{R}^5$ is the configuration space, the complete dynamics within a domain depends simultaneously on the Lagrangian as well as the contact constraints. All potential contacts of the WBRD with the floor (if not physical obstacles are considered) forces a holonomic constraint, $\eta_c(x_a)$. Considering that C_v is an indexing set of the possible holonomic constraints defined on D_v, then the holonomic constraints of the domain corresponds to $\eta_c(x_a) = \{\eta_c(x_a)\}_{c\in C_v}$ constant while the corresponding kinematic constraints corresponds to $J_v(x_a)\dot{q} = 0$, $J_v(q)$ is the Jacobian of $\eta_c(x_a)$, i.e., $J_v(x_a) = \dfrac{\partial\eta_c(x_a)}{\partial x_a}$.

The nature of the WBRD justifies that all the states (angular displacements and velocities) are uniformly bounded in time. Therefore, the state $x^\top = \left[x_a^\top, x_b^\top\right]$ is included in the set $X^+ = X_a^+ \cup X_b^+$ defined as:

$$\begin{aligned} X_a^+ &= \left\{x_a \mid -\infty < x_{a,i}^- \le x_{a,i} \le x_{a,i}^+ < +\infty, \right\} \\ X_b^+ &= \left\{x_b \mid -\infty < x_{b,i}^- \le x_{b,i} \le x_{b,i}^+ < +\infty, \right\} \end{aligned} \tag{4}$$

with x_i the i-th component of x, and the corresponding limits $\underset{t\ge 0}{sup}(x_i) + \epsilon = x_i^+$ and $\underset{t\ge 0}{inf}(x_i) - \epsilon = x_i^-$ with ϵ a small constant real scalar and x_i is either $x_{i,a}$ or $x_{i,b}$. Indeed, the set X^+ defines the holonomic restrictions for the WDRD structure.

Domains and Guards. A limited number of forces/moments appears if the holonomic constraints are active. These conditions can be represented in the form of component-wise inequalities $\nu_v F_v(x_a, x_b, u_v) \ge 0$, ν_v is the function containing the parameters of the WBRD. The unilateral constraints in the set U_v complete the admissible configurations for the WBRD via the domain of admissibility D_v:

$$D_v = \{(x_a, x_b, u_v) \in TQ \times U_v \mid A_v(x_a, x_b, u_v) \ge 0\} \tag{5}$$

for $v \in V$. The boundary of each sub-domain are characterized with

$$A_v(x_a, x_b, u_v) = \begin{bmatrix} \nu_v F_v(x_a, x_b, u_v) \\ \\ h_v(x_a) \end{bmatrix} \ge 0 \tag{6}$$

A state guard S_e corresponds to a proper subset of the domain D_v boundary, which is determined by an edge condition connected to the transition from D_v to the subsequent domain, D_v^+. Let us define $H_e(x_a; x_b; u_v)$ as an appropriate set of elements taken from (6) which characterize the edge condition. Using such elements, the guard can be characterized as

$$S_e = \{(x_a, x_b, u_v) \in TQ \times U_v \mid H_e(x_a; x_b; u_v) = 0; \quad \dot{H}_e(x_a; x_b; u_v) < 0\} \tag{7}$$

Discrete Dynamics. Consider the guard S_e as a reset map R_e that connects the system states over the guard to the subsequent domain. Considering the pre-impact states $(x_a^-; x_b^-)$ on S_e, the post-impact states $(x_a^+; x_b^+)$ of D_v^+ are computed using a reset map R_e by assuming the contact characterized by a perfectly plastic impact (if an impact occurs) [25]. Following the ideas in [26], the states configurations of the WBRD remain invariant during the impact, i.e., $(x_a^-; x_a^+)$; however, post-impact velocities must satisfy the plastic impact equation:

$$\begin{bmatrix} D(x_a^-) & -J_{v^+}^\top(x_a^-) \\ J_{v^+}^\top(x_a^-) & 0 \end{bmatrix} \cdot \begin{bmatrix} x_b^+ \\ \delta F_v \end{bmatrix} = \begin{bmatrix} D(x_a^-)x_b^- \\ 0 \end{bmatrix} \tag{8}$$

where δ defines the impulse function for the forces in the WBRD during the contact with the floor.

Virtual Constraints. Analogously to the described holonomic constraints, virtual constraints (recognized as the tracking errors in the control literature) correspond to the functions that modulates the dynamics the WBRD to track certain reference trajectories. The term *virtual* arises from the fact that such operative constraints must be enforced via a set of feedback (state or output) control instead of using forced physical restrictions. In equivalence to tracking errors, virtual constraints correspond to $\Delta_{s(t)}$. In here, the desired trajectories are proposed accordingly to the technique proposed in [24], where a novel technique to design monotonic and differentiable trajectories over a gait cycle is precisely detailed [27].

Here, one may notice that the goal of the proposed controller is steering $\Delta_{s(t)}$ to the origin if possible or at least to the zone (indeed, an invariant set) defined in (1) within each continuous domain. In this study, we avoid driving $\Delta_{s(t)}$ to the invariant set tracking through discrete dynamics. Considering that WBRD must realize movements with the aim of attaining the next switching configurations, the sequence of desired movements $x_a^* \in \mathbb{R}^5$ and $x_b^* \in \mathbb{R}^5$ should be calculated considering the distance between objects, the stable configurations for the WBRD, and so on. Notice that the position of the j-th articulation $x_{a,j}^*$ is known in advance assuming that the desired velocity $x_{b,j}^*$ can be estimated by direct differentiation (notice that the design of reference trajectories provides differentiable with continuous derivative flows).

Once the conditions to describe the WBRD have been detailed, it is feasible to propose the controller that can steer the virtual constraints to the origin or at least to the invariant set in (1).

5. Controller Design

5.1. Abstracted Representation of the WBRD

The aim of this research work is developing an output feedback loop controller for a WBRD, which should take into account the hybrid nature of the gait cycle and the state restrictions which define the angular restrictions at each joint. The proposed controller considers then the joints restrictions formed during the standing stage of the WBRD. The dynamics of the WBRD (considering the hybrid nature) is described as follows:

$$\begin{aligned} \frac{d}{dt}x_a(t) &= x_{b,s(t)}(t) \\ \frac{d}{dt}x_b(t) &= f_{s(t)}(x(t)) + g_{s(t)}(x_a(t))\,u(t) + \xi_{s(t)}(x(t),t) \end{aligned} \tag{9}$$

Here, $x_a \in \mathbb{R}^5$ is the vector of angular displacements of the joints considered in the WBRD. The vector $x_b \in \mathbb{R}^5$ is the vector of the angular velocities of all joints. The nature of WBRD structure enforces the existence of restrictions for all components in the state vector that is (4).

The function $f_{s(t)} : \mathbb{R}^{10} \times \mathbb{R}^+ \to \mathbb{R}^5$ in (9) represents the drift term that corresponds to internal dynamics of the BIMR:

$$\begin{aligned} \|f_{s(t)}(x^1) - f_{s(t)}(x^2)\| &\le L_f\|x^1 - x^2\|, \\ x^1 \in \mathbb{R}^{10}, x^2 &\in \mathbb{R}^{10}, L_f \in \mathbb{R}^+ \end{aligned} \tag{10}$$

The function $g_{s(t)} : \mathbb{R}^5 \to \mathbb{R}^5$ characterizes how the input function affects the robot dynamics. This function is invertible by the nature of the biped robot (formed as class of alternated robotic muti-articulated arm) and satisfies

$$0 < g^- \leq |g_{s(t)}(x_a)| \leq g^+ < +\infty \quad g^- \in \mathbb{R}^+, g^+ \in \mathbb{R}^+ \tag{11}$$

The bounded function $u \in \mathbb{R}^5$ is referred to as the control function, which must take into account the hybrid nature of the WBRD dynamics. By assumption, all the admissible controls belong to the following so-called admissible set:

$$U_{adm}=\left\{u : \|u\|^2 \leq u_0 + u_1\|x\|^2, u_0 \in \mathbb{R}^+, u_1 \in \mathbb{R}^+\right\} \tag{12}$$

The term $\xi_{s(t)} : \mathbb{R}^{10} \times \mathbb{R}^+ \to \mathbb{R}^5$ corresponds to admissible class of uncertainties and perturbations affecting the dynamics of WBRD. By assumption, the term $\xi_{s(t)}$ satisfies the following restriction:

$$\|\xi_{s(t)}(x,t)\|^2 \leq \xi_0 + \xi_1\|x\|^2, \;\; \xi_0 \in \mathbb{R}^+, \xi_1 \in \mathbb{R}^+ \tag{13}$$

5.2. H-ADRC Design

Considering the hybrid nature of the WBRD and the state restrictions, there are a few possible controllers that can be used. This study considers the application of a class of output feedback hybrid ADRC which can take into account the state constraints.

The design of the proposed H-ADRC considers the design of an approximation for the uncertain section of the WBRD which is valid within each continuous domain. In this study, let assume that the control free right-hand section of the WBRD dynamics ($F_{s(t)} = f_{s(t)} + \xi_{s(t)}$) can be represented as the composition of a nominal model $f_{0,s(t)}(x)$ added with a modeling function $\tilde{f}_{s(t)}(x,t)$, which represents those dynamical behaviors that are not modeled, which is $F_{s(t)}(x,t) = f_{0,s(t)}(x) + \tilde{f}_{s(t)}(x,t)$.

In this case, this uncertain section added to the external disturbances element can be represented as $\tilde{f}_{s(t)}(x,t) + \xi_{s(t)}(x(t),t)$, where $\tilde{f}_{s(t)}(x,t)$ represents the modeling error $\tilde{f}_{s(t)}(x,t) = f_{s(t)}(x,t) - f_{0,s(t)}(x)$ with $f_{0,s(t)} : \mathbb{R}^{10} \to \mathbb{R}^5$ describing the nominal model of the WBRD that could be estimated by diverse methods in such a way that the Euler–Lagrange modeling technique is still applicable. In this study, the first option is considered. Consequently, consider the following necessary assumption which must be used in the design of the H-ADRC.

Assumption 1: There exists a matrix of constants for each continuous subsystem $a_{s(t)} \in \mathbb{R}^{(p+1)\times 5}$ such that the function $F_{s(t)}$ evaluated over the trajectories $x = x(t)$ could be represented as $F_{s(t)}(x,t) = a_{s(t)}^{\top}\kappa(x) + \bar{f}_{s(t)}(x,t)$.

In this study, the time-dependent vector $\kappa \in \mathbb{R}^{p+1}$ (see [28,29] for further details) is

$$\kappa = [1, t, \cdots, t^p] \tag{14}$$

The term $\bar{f}_{s(t)}(x,t)$ is called the modeling error produced by the approximation of $F_{s(t)}(x,t)$ by a finite number p of elements in the basis and admits the following bounds by assumption

$$\|\bar{f}_{s(t)}(x,t)\| \leq f_0^+, \quad \forall t \geq 0 \tag{15}$$

The so-called nominal model for each continuous domain $a_{s(t)}^{\top}\kappa(t)$ can be expressed as $a_{s(t)}^{\top}\kappa(t) = a_{0,s(t)} + a_{1,s(t)}t + a_{2,s(t)}t^2 + \cdots + a_{p,s(t)}t^p$ [30]. In this study, the function $a_{s(t)}^{\top}\kappa(x(t))$ can be represented as a chain of integrators of some predefined constant matrices. Thus, the approximation presented above states that $F_{s(t)}(x,t)$ must be the solution of an integration operation of an uncertain function plus the approximation error: that is,

$$\begin{aligned} a_{s(t)}^{\top}\kappa(t) = a_{0,s(t)} + \int_{\tau_1=0}^{t} a_{1,s(t)} d\tau_1 + \int_{\tau_1=0}^{t}\int_{\tau_2=0}^{\tau_1} 2a_{2,s(t)} d\tau_2 d\tau_1 \\ + \cdots + \int_{\tau_1=0}^{t}\cdots\int_{\tau_p=0}^{\tau_{p-1}} p! a_{p,s(t)} d\tau_p \cdots d\tau_1 \end{aligned} \tag{16}$$

Equation (16) can be reorganized in an equivalent differential form:

$$\begin{aligned} a_{s(t)}^{\top}\kappa(t) = \rho_{0,s(t)}(t) \;\; \rho_{0,s(t)} = D^{\top}\rho_{s(t)}, \;\; D = \left[I_5, 0, \cdots, 0\right]^{\top}, D \in \mathbb{R}^{5*(p+1)} \\ \frac{d\rho_{s(t)}(t)}{dt} = \Phi\rho_{s(t)}(t), \;\; \Phi = \begin{cases} 1 & if \quad i = j-1 \\ 0 & if \quad otherwise \end{cases} \end{aligned} \tag{17}$$

The vector of initial conditions for $\rho_{s(t)}$ is $\rho_{s(t)}(0) = [a_{0,s(t)}, a_{1,s(t)}, a_{2,s(t)}, ..., a_{p,s(t)}]$. Now, the problem formulation given can be rephrased as follows: Given an output reference trajectory x^* for the system (9), let us design an output feedback controller that, regardless of the unknown non-modeled dynamics or external disturbances that forces the states x to track asymptotically the desired reference trajectories, with the tracking error restricted to a small neighborhood near the origin and proportional to a power of the uncertainties and perturbations. The first stage in solving this problem is designing an extended state observer to reconstruct the non-measurable part of the state.

5.3. Closed-Loop Dynamics Based on the H-ADRC Structure and Extended State Observer

Let us consider the reference trajectories x_a^* and x_b^* that are governed by

$$\begin{aligned} \frac{d}{dt}x_a^*(t) = x_{b,s(t)}^*(t) \\ \frac{d}{dt}x_b^*(t) = h_{s(t)}^*(t) \end{aligned} \tag{18}$$

where $h_{s(t)}^* : \mathbb{R}^+ \to \mathbb{R}^5$ is a continuous function with respect to time which can vary according to the active semi-cycle of the WBRD. The proposed reference trajectories satisfy the following bound for $x^* = [(x_a^*)^{\top}, (x_b^*)^{\top}]^{\top}$, $\|x^*\|^2 \leq x^{*,+}$, $x^{*,+} > 0$.

Based on the approximation proposed for $F_{s(t)}(x,t)$ and the reference trajectories given in (18), the dynamics of the tracking error Δ are given by

$$\begin{aligned} \frac{d}{dt}\Delta_a(t) = \Delta_{b,s(t)}(t) \\ \frac{d}{dt}\Delta_b(t) = D^{\top}\rho_{s(t)}(t) + \tilde{f}_{s(t)}(x(t),t) + \xi_{s(t)}(x(t),t) + g_{s(t)}(x_a(t))\,u(t) - h_{s(t)}^*(t) \\ \frac{d}{dt}\rho_{s(t)}(t) = \Phi\rho_{s(t)}(t) \end{aligned} \tag{19}$$

Notice that the bounds for the state x presented as holonomic constraints and the bounds for the reference trajectories provide the following estimation for the bounds of the tracking error Δ in each continuous domain:

$$\begin{aligned} \|\Delta(t)\|^2_{H_{s(t)}} &< V^+, V^+ = 2\lambda_{max}\left\{H_{s(t)}\right\}(x^+ + x^{*,+}) + \epsilon \\ x^+ &= \sum_{i=1}^{5}\left(\max\left\{x^-_{a,i}, x^+_{a,i}\right\}\right)^2 + \left(\max\left\{x^-_{b,i}, x^+_{b,i}\right\}\right)^2 \qquad \epsilon > 0 \end{aligned} \tag{20}$$

The design of the output feedback controller needs to provide an extended state robust state estimator of (9) which in this case satisfies the following hybrid dynamics:

$$\begin{aligned} \frac{d}{dt}\hat{x}_a(t) &= \hat{x}_{b,s(t)}(t) + L_{a,s(t)}e_{a,s(t)}(t) \\ \frac{d}{dt}\hat{x}_b(t) &= D^\top\hat{\rho}_{s(t)}(t) + g_{s(t)}\left(x_{a,s(t)}(t)\right)u(t) + L_{b,s(t)}e_{a,s(t)}(t) \\ \frac{d}{dt}\hat{\rho}_{s(t)}(t) &= \Phi\hat{\rho}_{s(t)}(t) + L_{c,s(t)}e_{a,s(t)}(t) \\ e_{a,s(t)}(t) &= \hat{x}_{a,s(t)}(t) - x_{a,s(t)}(t) \end{aligned} \tag{21}$$

Notice here that $e_{a,s(t)}(t) = C^\top(\hat{x}_{s(t)}(t) - x_{s(t)}(t))$, $\hat{x}_{s(t)} = [\hat{x}^\top_{a,s(t)}, \hat{x}^\top_{b,s(t)}]^\top$ and $C = [I_5, 0_5]^\top$. The observer gains are defined by $L_{a,s(t)} \in \mathbb{R}^{5\times5}$ and $L_{a,s(t)} \in \mathbb{R}^{5\times5}$. These gains must be calculated depending on what the active WBRD semi-cycle is.

The dynamics of $e_{a,s(t)}$ are associated with an extended state observer connected to:

$$\begin{aligned} \frac{d}{dt}e_a(t) &= e_{b,s(t)}(t) + L_{a,s(t)}e_{a,s(t)}(t) \\ \frac{d}{dt}e_b(t) &= D^\top\tilde{\rho}_{s(t)}(t) + L_{b,s(t)}e_{a,s(t)}(t) \\ \frac{d}{dt}\tilde{\rho}_{s(t)}(t) &= \Phi\tilde{\rho}_{s(t)}(t) + L_{c,s(t)}e_{a,s(t)}(t) \end{aligned} \tag{22}$$

where $\tilde{\rho}_{s(t)} = \hat{\rho}_{s(t)} - \rho_{s(t)}$.

Let us consider the proposed output-based controller satisfying:

$$u(t) = g^{-1}_{s(t)}\left(x_{a,s(t)}(t)\right)\left[-K^\top_{a,s(t)}\Delta_{a,s(t)} - K^\top_{b,s(t)}(\hat{x}_{b,s(t)}(t) - x^*_{b,s(t)}(t)) + h^*_{s(t)}(t) - D^\top\hat{\rho}_{s(t)}(t)\right] \tag{23}$$

where $K^\top_{a,s(t)} \in \mathbb{R}^{5\times5}$ and $K^\top_{b,s(t)} \in \mathbb{R}^{5\times5}$ are the piece-wise constant gains of the controller which are adjusted in each continuous dynamics.

Let us introduce the extended state vector $z \in \mathbb{R}^{10+10+5(p+1)}$ defined as $z = [\Delta^\top, e^\top, \tilde{\rho}^\top]^\top$ with $e = [e^\top_a, e^\top_b]^\top$. The dynamics of z are described by

$$\frac{d}{dt}z(t) = \Pi(K_{s(t)}, L_{es,s(t)}, L_{c,s(t)})z(t) + \Xi_{s(t)}(x(t), t) \tag{24}$$

with

$$
\begin{aligned}
\Pi(K_{s(t)}, L_{es,s(t)}, L_{c,s(t)}) &= \begin{bmatrix} A_{K,s(t)} & -BK_{b,s(t)}E & -BD^{\top} \\ 0_5 & A_{L,s(t)} & BD \\ 0_5 & L_{c,s(t)}C^{\top} & \Phi \end{bmatrix} \\
\Xi_{s(t)}(x,t) &= \begin{bmatrix} B\tilde{f}_{s(t)}(x,t) + B\xi_{s(t)}(x(t),t) \\ B\tilde{f}_{s(t)}(x,t) + B\xi_{s(t)}(x(t),t) \\ 0_{5(p+1)} \end{bmatrix}
\end{aligned} \tag{25}
$$

where $A_{K,s(t)} = A - BK_{s(t)}$, $A_{L,s(t)} = A + L_{es,s(t)}C^{\top}$, $\Lambda = \Lambda^{\top} \in \mathbb{R}^{N_p \times N_p}$, $B^{\top} = [0_5, I_5]$; and $E = [0_5, I_5]$.

The stability analysis considers the study over the dynamics of z. This analysis provides the result of the tracking controller, the state estimator, and the reconstruction of the uncertain section in the WBRD dynamics. This methodology yields the satisfaction of the close-loop analysis of the output feedback controller which offers a class of separation-principle for the proposed design. This is an additional theoretical contribution of this study. The following theorem details the main result of this study.

Theorem 1. *Consider the state observer given in (21) and the output feedback controller proposed in (23) with gains adjusted such that all matrices $A - BK_{s(t)}$ and $A + L_{es,s(t)}C^{\top}$ are Hurwitz for the WBRD dynamics with incomplete information approximated with (17).*

If there is a sequence of positive definite matrices $Q_{R,s(t),T_\kappa}$ and $Q_{L,s(t),T_\kappa}$ such that positive definite and symmetric solutions $H_{s(t),T_\kappa} > 0$ and $M_{s(t),T_\kappa}$ exist for the following matrix inequalities $Ric_{s(t)}(H_{s(t),T_\kappa}) \leq 0$, $Lyap_{s(t)}(M_{s(t),T_\kappa}) \leq 0$ and $Lyap^{D}_{s(t)}N_{s(t),T_\kappa}) \leq 0$ with

$$
Ric_{s(t)}(H_{s(t),T_\kappa}) = H_{s(t),T_\kappa}A_{K,s(t)} + A^{\top}_{K,s(t)}H_{s(t),T_\kappa} + H_{s(t),T_\kappa}(\Lambda_1 + \Lambda_2)H_{s(t),T_\kappa} + Q_{R,s(t),T_\kappa} \tag{26}
$$

$$
Lyap_{s(t)}(M_{s(t),T_\kappa}) = M_{s(t),T_\kappa}\Pi_{2,s(t)} + \Pi^{\top}_{2,s(t)}M_{s(t),T_\kappa} + \Pi^{\top}_{2,s(t)}\Lambda_3\Pi_{2,s(t)} + \epsilon^{-1}\Pi^{\top}_{1,s(t)}\Lambda_1^{-1}\Pi_{1,s(t)} + Q_{L,s(t),T_\kappa} \tag{27}
$$

$$
\begin{aligned}
Lyap^{D}_{s(t)}N_{s(t),T_\kappa}) &= \begin{bmatrix} \Pi^{\top}(T_\kappa)N_{s(t),T_{\kappa+1}}\Pi(T_\kappa) - N_{s(t),T_\kappa} & \Pi^{\top}(T_\kappa)\Xi_{s(t)}(x(T_\kappa),T_\kappa) \\ \Xi^{\top}_{s(t)}(x(T_\kappa),T_\kappa)\Pi(T_\kappa) & N_{k,T_{\kappa+1}} \end{bmatrix} \\
\Pi(T_\kappa) &= \Pi(K_{s(T_\kappa)}, L_{es,s(T_\kappa)}, L_{c,s(T_\kappa)})
\end{aligned} \tag{28}
$$

then the extended state z converges exponentially to the invariant set $\mathcal{I}_{D,T_\kappa} \times \mathcal{I}_{Z,T_\kappa}$ defined by

$$
\mathcal{I}_{D,T_\kappa} \times \mathcal{I}_{Z,T_\kappa} = \left\{ (\Delta, z_0) \quad | \quad \|\Delta\|^2_{Q_{k,T_\kappa} - Q_{k,T_\kappa,0}} \geq \beta_2, \quad \|z_0\|^2_{Q_{L,k,T_\kappa} - Q_{L,k,T_\kappa,0}} \geq \beta_2 \right\}
$$

with $\beta_j = 2\lambda_{max}\left(B^{\top}\Lambda_j^{-1}B\right)\left(\xi_0 + \xi_1(x^+)^2 + f_0^+\right)$, $j = 2,3$, $\Lambda_2 \in \mathbb{R}^{5\times 5}$, $\Lambda_3 \in \mathbb{R}^{5\times 5}$ are positive and symmetric definite matrices and $Q_{k,T_\kappa,0} \in \mathbb{R}^{5\times 5}$, $Q_{L,k,T_\kappa,0} \in \mathbb{R}^{(5+5(p+1))\times(5+5(p+1))}$ are positive definite matrices fulfilling $Q_{k,T_\kappa} > Q_{L,k,T_\kappa,0}$, $Q_{k,T_\kappa} > Q_{L,k,T_\kappa,0}$.

The rate of exponential convergence is given by:

$$
\alpha_k = min\left\{\lambda_{min}\left\{H_k^{-1/2}Q_{k,0}H_k^{-1/2}\right\}, \lambda_{min}\left\{M_k^{-1/2}Q_{L,k,0}M_k^{-1/2}\right\}\right\}
$$

Proof. The estimation of the adjustment laws for the controller gains uses the concept of Lyapunov stability based on BLF. Formally, the proposed BLF which is used to prove the stability of the origin considers as a class of practical equilibrium point for the movement of the WBRD. In this study, the logarithmic function is used, one of the most common BLFs [31,32]. The suggested BLF function to get the stability analysis in this study is given by

$$V(z) := \sum_{k=1}^{2} V_{k,T_\kappa}(\Delta, z_0)$$
$$V_k(\Delta, z_0) = \left[ln\left(\frac{V^+}{V^+ - \|\Delta\|^2_{H_{k,T_\kappa}}}\right) + z_0^\top M_{k,T_\kappa} z_0 \right], \quad z_0 = [e^\top, \tilde{\rho}^\top]^\top \tag{29}$$

where $k = 1, 2$. Notice then that $s(t) = 1$ represents the case a and $s(t) = 2$ represents the case b.

The full-time derivative of $V(z)$ is

$$\frac{d}{dt}V_{k,T_\kappa}(\Delta(t), z_0(t)) = \left[\frac{2\Delta^\top(t) H_{k,T_\kappa}}{V^+ - \|\Delta(t)\|^2_{H_{k,T_\kappa}}} \frac{d}{dt}\Delta(t) + 2z_0^\top(t) M_{k,T_\kappa} \frac{d}{dt} z_0(t) \right] \tag{30}$$

Reorganizing the differential equation (30) yields

$$\frac{d}{dt}V_k(\Delta(t), z_0(t)) = 2\left[\frac{\Delta^\top(t) H_{k,T_\kappa}}{V^+ - \|\Delta(t)\|^2_{H_{k,T_\kappa}}}, \quad z_0^\top(t) M_{k,T_\kappa} \right] \frac{d}{dt} z(t) \tag{31}$$

The substitution of $\frac{d}{dt}z(t)$ on the full-time derivative of $V(z(t))$ leads to the following form:

$$\frac{d}{dt}V_k(\Delta(t), z_0(t)) = 2\left[\frac{\Delta^\top(t) H_{k,T_\kappa}}{V^+ - \|\Delta(t)\|^2_{H_{k,T_\kappa}}}, \quad z_0^\top(t) M_{k,T_\kappa} \right] \left[\Pi(K_k, L_{es,k}, L_{c,k}) z(t) + \Xi_k(x(t), t) \right] \tag{32}$$

Notice that the term $\Delta^\top(t) H_{k,T_\kappa} \Pi(K_k, L_{es,k}, L_{c,k}) z(t)$ can be handled as follows:

$$\Delta^\top(t) H_{k,T_\kappa} \Pi(K_k, L_{es,k}, L_{c,k}) z(t) = \Delta^\top(t) H_{k,T_\kappa} (A - BK_k)\Delta(t) + \Delta^\top(t) H_{k,T_\kappa} \Pi_{1,k} z_0(t) \tag{33}$$

where $\Pi_{1,k} = \begin{bmatrix} -BK_{b,k}E & -BD^\top \end{bmatrix}$.

Let us consider the application of the Young inequality, which satisfies:

$$X^\top Y + Y^\top X \leq X^\top N X + Y^\top N^{-1} Y$$

valid for any $X, Y \in \mathbb{R}^{r \times s}$ and any $0 < N = N^\top \in \mathbb{R}^{s \times s}$ [33]. Therefore, the following upper bound for $\Delta^\top(t) H_k \Pi_{1,k} z_0(t)$ is valid:

$$2\Delta^\top(t) H_{k,T_\kappa} \Pi_{1,k} z_0(t) \leq \|H_{k,T_\kappa}\Delta(t)\|^2_{\Lambda_1} + \|\Pi_{1,k} z_0(t)\|^2_{\Lambda_1^{-1}} \tag{34}$$

In equivalent form, $\Delta^\top(t) H_{k,T_\kappa} \Xi_k(x(t), t)$ accepts the following upper bound:

$$\Delta^\top(t) H_{k,T_\kappa} B\tilde{f}_k(x,t) \leq \|H_{k,T_\kappa}\Delta(t)\|^2_{\Lambda_2} + \|B\tilde{f}_k(x,t) + B\xi_k(x(t),t)\|^2_{\Lambda_2^{-1}}, \tag{35}$$

Introducing the following matrix Π_2

$$\Pi_2 = \begin{bmatrix} 0_5 & I_5 & 0_{5(p+1)} \\ 0_5 & 0_5 & I_{5(p+1)} \end{bmatrix} \times \begin{bmatrix} A_{K,k} & -BK_{b,k}E & -BD^\top \\ 0_5 & A_{L,k} & B\,D \\ 0_5 & L_{c,k}C^\top & \Phi \end{bmatrix} = \begin{bmatrix} A_{L,k} & B\,D \\ L_{c,k}C^\top & \Phi \end{bmatrix}$$

The terms including $z_0^\top(t)M_{k,T_\kappa}$ in the time derivative of $V(z(t))$ can be presented as

$$
\begin{aligned}
&2z_0^\top(t)M_{k,T_\kappa}\Pi_{2,k}z_0(t) + 2z_0^\top(t)M_{k,T_\kappa}\Pi_{2,k}\Xi_{s(t)}(x(t),t) \leq \\
&z_0^\top(t)\left(M_{k,T_\kappa}\Pi_{2,k} + \Pi_{2,k}^\top M_{k,T_\kappa}\right)z_0(t) + \|\Pi_{2,k}z_0(t)\|^2_{\Lambda_3} + \|B\tilde{f}_k(x,t) + B\xi_k(x(t),t)\|^2_{\Lambda_3^{-1}}
\end{aligned}
\tag{36}
$$

Taking together the results in (33) to (36) yields

$$
\begin{aligned}
\frac{d}{dt}V_k(\Delta(t),z_0(t)) \leq &\left[\frac{2\Delta^\top(t)H_{k,T_\kappa}A_{K,k}\Delta(t) + \|H_{k,T_\kappa}\Delta(t)\|^2_{\Lambda_1+\Lambda_2} + \|B(\tilde{f}_k(x,t)+\xi_k(x(t),t))\|^2_{\Lambda_2^{-1}} + \|\Pi_1 z_0(t)\|^2_{\Lambda_1^{-1}}}{V^+ - \|\Delta(t)\|^2_{H_{k,T_\kappa}}}\right] + \\
&z_0^\top(t)\left(M_{k,T_\kappa}\Pi_{2,k} + \Pi_{2,k}^\top M_{k,T_\kappa}\right)z_0(t) + \|\Pi_{2,k}z_0(t)\|^2_{\Lambda_3} + \|B(\tilde{f}_k(x,t)+\xi_k(x(t),t))\|^2_{\Lambda_3^{-1}}
\end{aligned}
\tag{37}
$$

Noticing that $min_\Delta\left\{V^+ - \|\Delta(t)\|^2_{H_{k,T_\kappa}}\right\} = \epsilon$, then

$$
\begin{aligned}
\frac{d}{dt}V_k(\Delta(t),z_0(t)) \leq &\left[\frac{\Delta^\top(t)\left(H_{k,T_\kappa}A_{K,k} + A_{K,k}^\top H_{k,T_\kappa} + H_{k,T_\kappa}(\Lambda_1+\Lambda_2)H_{k,T_\kappa}\right)\Delta(t) + \|B(\tilde{f}_k(x,t)+\xi_k(x(t),t))\|^2_{\Lambda_2^{-1}}}{V^+ - \|\Delta(t)\|^2_{H_{k,T_\kappa}}}\right] + \\
&z_0^\top(t)\left(M_{k,T_\kappa}\Pi_{2,k} + \Pi_{2,k}^\top M_{k,T_\kappa} + \Pi_{2,k}^\top\Lambda_3\Pi_{2,k} + \epsilon^{-1}\Pi_{1,k}^\top\Lambda_1^{-1}\Pi_{1,k}\right)z_0(t) + \|B(\tilde{f}_k(x,t)+\xi_k(x(t),t))\|^2_{\Lambda_3^{-1}}
\end{aligned}
\tag{38}
$$

Based on the upper bounds of (13) and the bounds for the modeling error yields to estimating upper norms of $\|B(\tilde{f}_k(x,t)+\xi_k(x(t),t))\|^2_{\Lambda_2^{-1}}$ and $\|B(\tilde{f}_k(x,t)+\xi_k(x(t),t))\|^2_{\Lambda_3^{-1}}$ as:

$$
\|B(\tilde{f}_k(x,t)+\xi_k(x(t),t))\|^2_{\Lambda_j^{-1}} \leq \beta_j, \tag{39}
$$

Similarly, the time derivative of (38) can be bounded as

$$
\begin{aligned}
\frac{d}{dt}V_k(\Delta(t),z_0(t)) \leq &\left[\frac{\Delta^\top(t)\left(H_{k,T_\kappa}A_{K,k} + A_{K,k}^\top H_{k,T_\kappa} + H_k(\Lambda_1+\Lambda_2)H_k\right)\Delta(t) + \beta_2}{V^+ - \|\Delta(t)\|^2_{H_{k,T_\kappa}}}\right] + \\
&z_0^\top(t)\left(M_{k,T_\kappa}\Pi_{2,k} + \Pi_{2,k}^\top M_{k,T_\kappa} + \Pi_{2,k}^\top\Lambda_3\Pi_{2,k} + \epsilon^{-1}\Pi_{1,k}^\top\Lambda_1^{-1}\Pi_{1,k}\right)z_0(t) + \beta_3
\end{aligned}
\tag{40}
$$

Notice that (40) can be represented as follows:

$$
\begin{aligned}
\frac{d}{dt}V_k(\Delta(t),z_0(t)) \leq &\left[\frac{\Delta^\top(t)Ric_k(H_{k,T_\kappa})\Delta(t) - \Delta^\top(t)Q_{k,T_\kappa}\Delta(t) + \beta_2}{V^+ - \|\Delta(t)\|^2_{H_{k,T_\kappa}}}\right] + \\
&[z_0^\top(t)Lyap_k(M_{k,T_\kappa})z_0(t) - z_0^\top(t)Q_{L,k,T_\kappa}z_0(t) + \beta_3]
\end{aligned}
\tag{41}
$$

Taking into account the assumptions that $Ric_k(H_{k,T_\kappa}) < 0$ and $Lyap_k(M_{k,T_\kappa}) < 0$ yielding

$$
\frac{d}{dt}V_k(\Delta(t),z_0(t)) \leq -\left[\frac{\Delta^\top(t)Q_{k,T_\kappa}\Delta(t) - \beta_2}{V^+ - \|\Delta(t)\|^2_{H_{k,T_\kappa}}} + z_0^\top(t)Q_{L,k,T_\kappa}z_0(t) - \beta_3\right] \tag{42}
$$

If we consider that $\Delta \in \mathcal{I}_{D,T_\kappa}$ and $z_0 \in \mathcal{I}_{Z,T_\kappa}$, then

$$
\frac{d}{dt}V_k(\Delta(t),z_0(t)) \leq -\left[\frac{\Delta^\top(t)Q_{k,T_\kappa,0}\Delta(t)}{V^+ - \|\Delta(t)\|^2_{H_{k,T_\kappa}}} + z_0^\top(t)Q_{L,k,T_\kappa,0}z_0(t)\right] \tag{43}
$$

Following the ideas given in [20], it is possible to prove that

$$
\frac{d}{dt}V_k(\Delta(t),z_0(t)) \leq -\alpha_k V_k(\Delta(t),z_0(t)) \quad \forall(\Delta,z_0) \notin \mathcal{I}_{D,T_\kappa} \times \mathcal{I}_{Z,T_\kappa} \tag{44}
$$

Consequently, $V_k(\Delta(t), z_0(t))$ converges asymptotically to the invariant set $\mathcal{I}_{D,T_\kappa} \times \mathcal{I}_{Z,T_\kappa}$ within a given continuous sub-domain. This is enough to prove the stability within each continuous sub-domain. Now, to prove the stability of the hybrid form, let us consider that the tracking error is already bounded; then, let us propose the discrete analysis for the dynamics of z evaluated on the specific times where the sequential transition from a -> b or vice versa. With the aim of evaluating this stability analysis, one may propose the discrete Lyapunov-like function such as

$$V^d_{T_\kappa}(z(T_\kappa)) := \sum_{k=1}^{2} V^d_{k,T_\kappa}(z(T_\kappa)) \qquad V^d_{k,T_\kappa}(z(T_\kappa)) = z^\top(T_\kappa) N_{k,T_\kappa} z(T_\kappa)$$
$$N_{k,T_\kappa} = \begin{bmatrix} H_{k,T_\kappa} & 0_5 \\ 0_{5+5(p+1)} & M_{k,T_\kappa} \end{bmatrix} \tag{45}$$

The discrete analysis of the discrete Lyapunov like function yields

$$\Delta V^d_{k,T_\kappa}(z(T_\kappa)) = z^\top(T_\kappa) N_{k,T_{\kappa+1}} z(T_\kappa) - z^\top(T_\kappa) N_{k,T_\kappa} z(T_\kappa) \tag{46}$$

Notice that

$$\Delta V^d_{k,T_\kappa}(z(T_\kappa)) = z^\top(T_{\kappa+1}) N_{k,T_{\kappa+1}} z(T_{\kappa+1}) - z^\top(T_\kappa) N_{k,T_\kappa} z(T_\kappa)$$
$$= \begin{bmatrix} z^\top(T_\kappa) & \Xi_k^\top(x(T_\kappa), T_\kappa) \end{bmatrix} \begin{bmatrix} \Pi^\top(T_\kappa) N_{k,T_{\kappa+1}} \Pi(T_\kappa) - N_{k,T_\kappa} & \Pi^\top(T_\kappa)\Xi_k(x(T_\kappa), T_\kappa) \\ \Xi_k^\top(x(T_\kappa), T_\kappa)\Pi(T_\kappa) & N_{k,T_{\kappa+1}} \end{bmatrix} \begin{bmatrix} z(T_\kappa) \\ \Xi_k(x(T_\kappa), T_\kappa) \end{bmatrix} \tag{47}$$

With the assumption that the matrix inequality (28) is negative definite, then, $\Delta V^d_{k,T_\kappa}(z(T_\kappa))$ is negative and, therefore, the discrete jumps remain negative confirming the local asymptotically stability of the origin for the extended system based on the state z. □

Remark 1. *Notice that the H-ADRC controller can be useful if the proposed control gains can be sufficiently adequate such that $SW^* \leq \beta_2$. This fact can be guaranteed a priori if a formal optimization of the size for the invariant set proposed in the statement of Theorem 1. The solution of this aspect is outside the scope of this study. However, we assume that the condition described in this remark is fulfilled.*

Remark 2. *Notice that adjusting the gains in adaptive form could reduce the large amplitude oscillations along the transient period of the tracking trajectory process. The adaptive adjustment of the gains satisfies:*

$$\frac{d}{dt}\hat{K}_{s(t)} = -\left(V^+ - \|\Delta\|^2_{H_{k,T_\kappa}}\right)^{-1} \Omega_k^{-1} H_{K,k} \Delta_{s(t)} \tag{48}$$

with $\Omega \in \mathbb{R}^{5\times 5}$. This result can be obtained directly with a similar stability analysis to the one introduced in Theorem 1. The main change is introducing a modified Lyapunov like function satisfying

$$V^{ad}(z) := \sum_{k=1}^{2} V^{ad}_{k,T_\kappa}(\Delta, z_0)$$
$$V^{ad}_k(\Delta, z_0) = ln\left(\frac{V^+}{V^+ - \|\Delta\|^2_{H_{k,T_\kappa}}}\right) + z_0^\top M_{k,T_\kappa} z_0 + trace\left\{\tilde{K}^\top_{s(t)} \Omega_k \tilde{K}_{s(t)}\right\} \tag{49}$$

where $\tilde{K}_{s(t)} = \hat{K}_{s(t)} - K_{s(t)}$ and trace refers to the trace operator. A similar study analysis yields the design of the adaptive gains which can presumably reduce the oscillating transitions.

Remark 3. *The result attained above requires the design of the extended state observer (21), which must provide efficient approximation of the angular velocities of all the articulations. Such condition implies complex instrument requirements for the BIMR. Such condition enforces that the estimation error must converge faster to the corresponding invariant set than the tracking controller does. A possible alternative is using some variant* *of robust time differentiator which can produce the estimation of the required angular velocities. The so-called super-twisting algorithm can provide such solution, but the close loop stability analysis requires some further work. The reader is referred to the studies given in [34] for more details.*

Remark 4. *The application presented in this manuscript needs to solve six different matrix inequalities offline. All of them are Riccati equations and their solutions are quite regular in control theory. Indeed, there exist numerical solvers that can help to find the solution of these inequalities. The requirements to find the solution of a Riccati equation (in general form) given by*

$$A^\top P + PA + PRP + Q = 0, \quad P = P^\top > 0 \tag{50}$$

are

- *The matrix A is Hurwitz, as a consequence:*
- *The pair $(A, R^{1/2})$ is controllable.*
- *The pair $(A, Q^{1/2})$ is observable.*

Notice that the stability of matrix A in our case is related with the gain matrices $K_{s(t)}$ and $L_{es,s(t)}$ that can be selected in such a way $A - BK_{s(t)}$ and $A - L_{es,s(t)}C^\top$ are Hurwitz.

6. Implementation Issues

The proposed H-ADRC controller requires several technical aspects that must be considered before it can be implemented. This section details the arrangement of all the aspects needed to realize both the numerical and the experimental evaluations.

6.1. Numerical Evaluation

In the simulation system, it is necessary to introduce a force sensing element. The information of the force sensor is used to detect the moment when the corresponding ending section of the WBRD has touched the surface. Notice that such contact must be part of the condition to switch between the subsystems that define the gait scenarios *a* and *b*. Including this additional sensor in the condition ensures that the WBRD is completing the semi-cycle in the adequate configuration.

In this study, the numerical simulations used a virtualized representation of the WBRD based on the SimMechanics Toolbox® of Matlab®. The virtual model includes all the articulations and the mechanical representation of vacuum pumps that are going to be used as the electro-mechanical elements to change the reference frame in the hybrid representation of the gait cycle of the WBRD and allowing for evaluating the suggested controller. The mechanical representation of the WBRD was prepared in the Solid-Works® software including all the mobile actions that must be exerted by the WBRD (Figure 5).

Figure 5. WBRD exported to simMechanics for simulation.

6.2. Experimental Evaluation

The experimental evaluation of the proposed controller was implemented in a polymer-based WBRD. The design of the robot followed the structure proposed in the numerical evaluation. The experimental prototype was constructed using the 3D printing technique using poly-lactic acid (PLA) as building material.

The constructed WBRD used DC motors to realize the mobilization of all joints. A set of gears transmits perpendicular movement to the mechanical structure to reach the desired angular trajectories. Each of the DC actuators was regulated with a DC source to alternate power converter using a pulse width modulation (PWM) methodology.

The numerical realization of the controller used a distributed strategy considering the combination of a processing board (TIVA1294 from Texas Instruments) and a personal computer (Alienware 17S from Dell Computers). The processing board realizes the PWM formulation based on the calculated control action in the personal computer (PC).

The PC realizes an image-based-processing algorithm which calculates the articulation angles using physical markers placed over the WBRD structure (Figure 6). The algorithm is described in Algorithm 1 which is based on the application of simplified morphological image processing methods. The first algorithm is complemented with the calculus of the state estimator (21) and then the output feedback controller proposed in (23) is evaluated according to Algorithm 2. The estimated control action is sent to the processing board via a serial protocol (RS-232).

The H-ADRC requires including vacuum pumps at the first and the last links of the bio-inspired robot. Each of the pumps is activated once all the angles have attained their reference values. The pump is activated to define the change of the reference framework. The activation action is also evaluated in the algorithm and then sent to the processing board.

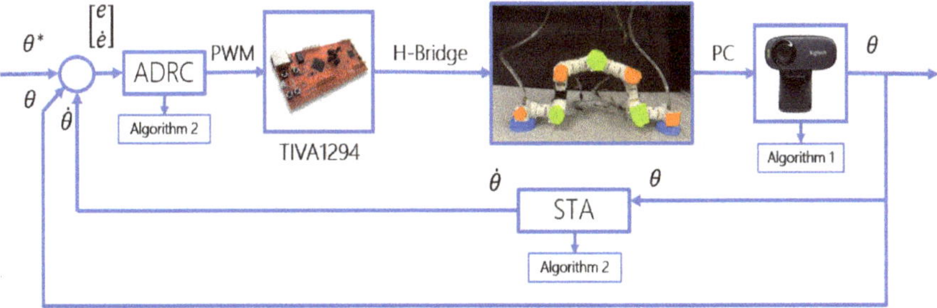

Figure 6. Implementation control in the WBRD.

Algorithm 1 Position image recovery

Start
$P \leftarrow$ a frame form the camera ▷ The threshold is selected to find the marker place on the robot joints
if $P_{i,j,3} < 10$ and $P_{i,j,1} > 180$ **then** ▷ Detect orange markers
    $A_{i,j} = 1$
else
    $A_{i,j} = 0$
end if
if $P_{i,j,3} < 60$ and $P_{i,j,1} < 128$ **then** ▷ Detect green markers
    $B_{i,j} = 1$
else
    $B_{i,j} = 0$
end if
$A = imclose(A)$ ▷ Morphological closing of the image
$B = imopen(B)$ ▷ Morphological opening of the image
$G = A + B$ ▷ Create logical image
$C = DetectCentroids(G)$ ▷ Function to detect centroids
for $i = 1, 2, ..., 6$ **do** ▷ Calculating the absolute angle with the slope of neighbor centroids
    $T_i = \arctan(\frac{C_2^{i+1} - C_2^i}{C_1^{i+1} - C_1^i})$
end for
$\theta_1 = T_1$
$S = 0$
for $i = 1, 2, ..., 5$ **do** ▷ Calculating the relative angle, with respect with the first one
    $S = S + \theta_i$
    $\theta_{i+1} = T_i - S$
end for
Return θ
Start control calculation ▷ Go to Algorithm 2

Algorithm 2 Control implementation

Start
Obtain the corresponding angles $\theta_{j,i}$ ▷ From Algorithm 1
Check the switching condition ▷ To see what pump is active and the corresponding system $j = 1, 2$
Implement the extended observer to recover $\dot{\theta}_{j,i}$ from Equation (17) ▷ To recover $\hat{\dot{\theta}}_{j,i}$ and $\hat{\rho}_{j,i}$
Implement the control law in (23) for the corresponding system
Evaluate the obtained decoupled controllers to convert into a pulse modulation signal (PWM)
if $|u_{j,i}| > 255$ **then** ▷ These values correspond to the high time in the PWM signal
    $u_{j,i} = 255$
else
    $u_{j,i} = u_{j,i}$
end if
Evaluete the control to determine the movement direction of the actuators
if $\text{sign}(u_{j,i}) > 1$ **then**
    $d_{j,i} = 1$
else
    $d_{j,i} = 0$
end if
Send the values through serial comunication (RS-232 protocol) to the TIVA1294
Activate a PWM with the values of $u_{j,i}$ and $d_{j,1}$ to be sent to the H bidges
Return to Algorithm 1 to calculate again the current position

7. Simulation Results

The proposed output feedback controller was evaluated using a set of numerical evaluations considering the exerting of three completes gait cycles . The corresponding sequences of reference angular movements were calculated using a biomechanical study of a *Leptidoptera gonodonta* or measuring worm. Once the angles were calculated, the method to produce the reference trajectories was implemented. These reference trajectories were injected into the SimMechanics software.

The numerically simulated model in SimMechanics-Matlab was evaluated considering the real masses (assuming the construction based on PLA material) and dimensions of each mechanical section of the WBRD. This strategy allowed for evaluating the controller as well as tuning the gains of both the estimators and controllers. These gains were used as the initial values in the experimental device.

The angular trajectories measured from the simulated WBRD were compared with the reference trajectories (Figure 7). The shown trajectories correspond to the reference signals, the measured position with the proposed H-ADRC (using the estimated velocities from the state estimator), and the state feedback form. The comparison of all trajectories confirms that the H-ADRC controller provides an equally faster convergence than other controllers, but it has less oscillations during the transient period. Such characteristic is a consequence of the additional compensation provided by the extended state observer that can actively compensate the effect of external perturbations and internal modeling imprecision. The comparison with a classical PID controller confirms such additional benefit of introducing the augmented compensation aggregated in the H-ADRC form. In addition, the proposed controller tracks the reference with smaller deviations than all other controllers considered for comparison. Notice also that these trajectories confirm the presence of high-frequency oscillations at the beginning of the tracking period (first three seconds). Although these oscillations may be undesired for the WBRD movements, the tracking exerted after the oscillations period justifies the introduction of H-ADRC based compensation due to its robustness against matched perturbations.

Figure 8 shows (in logarithmic scale) the control associated energy enforced by the state feedback (marked with PD) and the H-ADRC controllers. This comparison considers that both controllers solve the tracking with the same convergence quality. The application of the compensated control form consumes smaller amounts of energy and augments the working life of the DC motors' actuators. These controllers were chosen for this comparison because they provided the best trajectory tracking among the evaluated controllers.

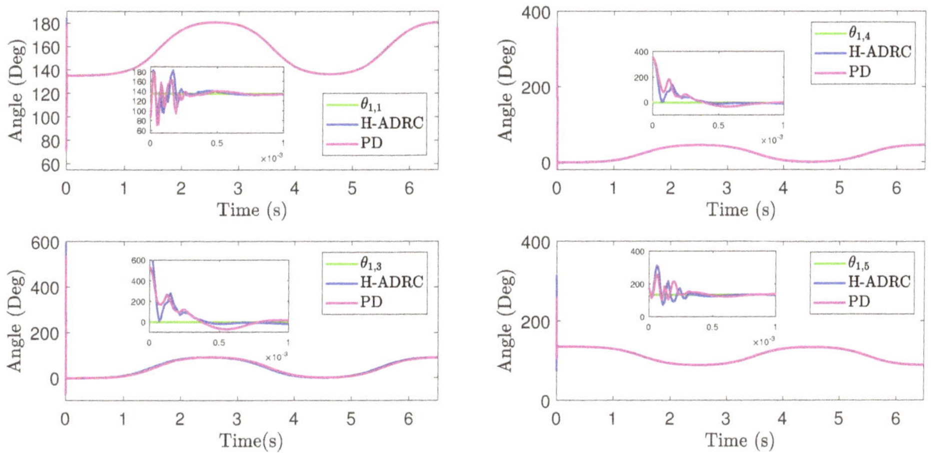

Figure 7. Comparison of reference trajectories with controlled states implementing either the PD and the proposed H-ADRC controllers.

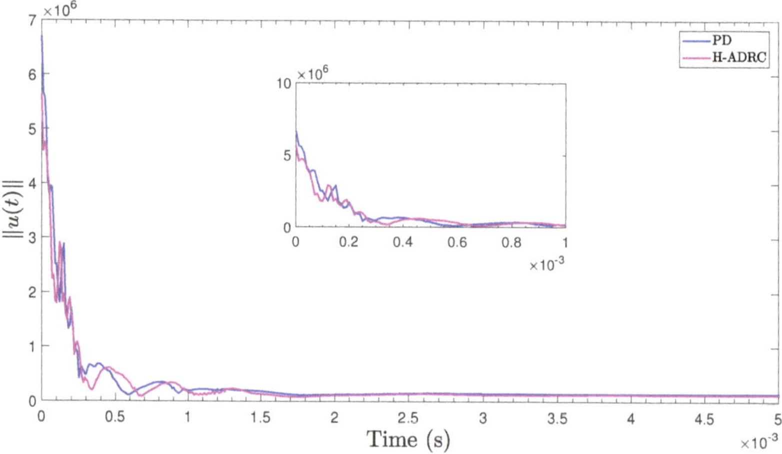

Figure 8. Euclidean norm of the control energy used with each control algorithm.

Figure 9 details the comparison of the Euclidean norms of the tracking errors obtained with the application of the same evaluated controllers that were presented in Figure 8. This figure highlights the rate of convergence and the ultimately bounded zone for the tracking errors. This figure is also presented in logarithm scale for the purpose of better detecting the differences among the proposed controllers. In both cases, the H-ADRC forces a faster convergence and smaller oscillations amplitude in steady-state for the controlled trajectories.

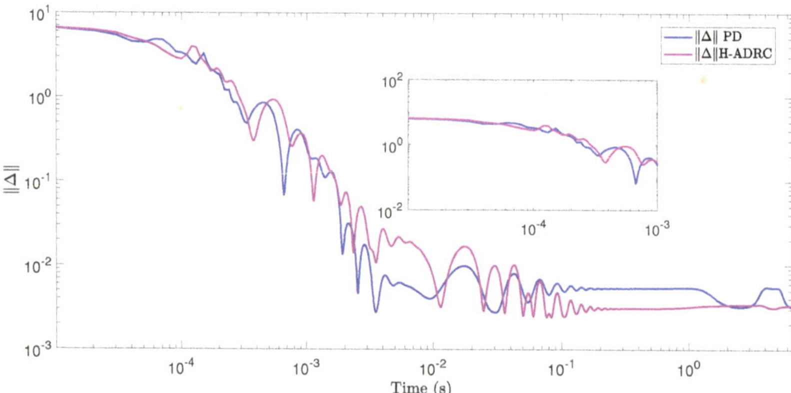

Figure 9. Euclidean norm of the tracking error for each control technique used in simulation.

Figure 10 demonstrates a sequence of image captures obtained from the simulated evaluation of the H-ADRC application over the WBRD. This sequence highlights the sequence of movements exerted by the entire simulated WBRD associated with the sequence of articular movements enforced by the distributed form of the proposed controllers. The sequence also demonstrates the benefits of introducing the simulated SimMechanics model because it allows for getting an efficient gains adjustment which yields to satisfying the complete gait sequence. Moreover, the hybrid analysis of the controller is confirmed with the efficient tracking of the reference angular positions in both scenarios a and b.

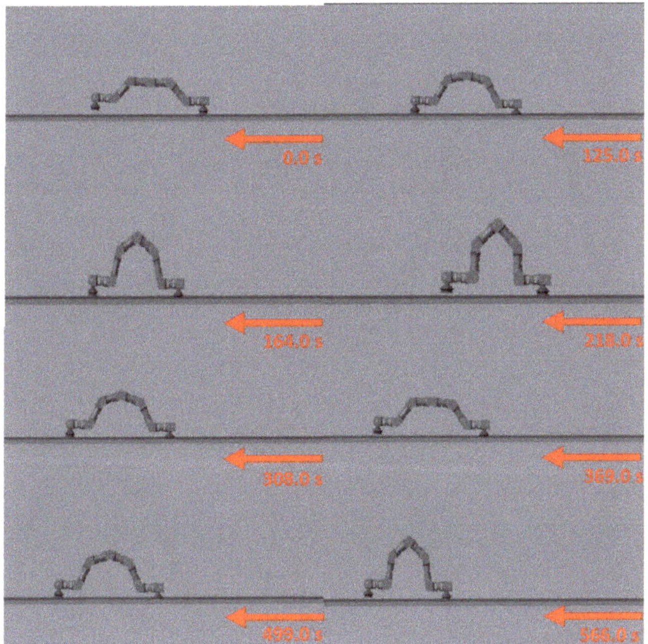

Figure 10. Sequence of movements of the gait cycle of the WBRD in simulation.

8. Experimental Results

Once the simulated evaluation showed acceptable results measured in terms of the tracking errors and the consumed energy, the control was implemented according to Algorithms 1 and 2. Different controllers were implemented with the aim of evaluating the advantages of the proposed methodology. The H-ADRC controller was compared with the classical state feedback controller using the derivative obtained by means of the extended state observer and an experimental PID form.

Figure 11 shows the comparison of the angular displacements obtained in the experimental results with three different controllers. For the PD controller supplied with the estimated derivative, the vector of the five different proportional gains were selected as $k_P = [23, 45, 50, 45, 23]$ and the derivative gains were $k_D = [2.41, 3.51, 4, 3.51, 2.41]$. The case of the integral part included the same proportional and derivative gains while the integral gains were: $k_I = [0.5, 0.9, 1.2, 0.75, 0.3]$. The hardware configuration used to evaluate the proposed controller provided an updating time of the control action of 0.05 s, which was enough to successfully realize the gait cycle by the experimental WBRD.

The comparison of the proposed controllers confirmed that the observed additional compensation of the H-ADRC improves the tracking efficiency for all the articulations. Moreover, the oscillations of the measured angular are reduced during the transient period. In addition, one may notice that state feedback provides the worst tracking performance among the evaluated controllers. This result was also true for all the trajectories in the constructed WBRD.

The comparison of the controllers' performances was realized through the calculus of the norm of the tracking errors (Figure 12a). This comparison proves that the tracking error is smaller if the evaluated controller was the H-ADRC in comparison with the other two controllers (state feedback and PID). Notice that the PID form provides a comparable tracking quality to the H-ADRC. Notice that a fair comparison between the evaluated controllers cannot include the norm of the tracking error only, but it must include the energy associated with the controller. Here, one may notice that H-ADRC uses larger energy (measured in terms of the norm of the control action) than the other two controllers

(Figure 12b). This increment is actually not significant (12%), and it occurs only during the 20% of the evaluated period corresponding to the gait cycle.

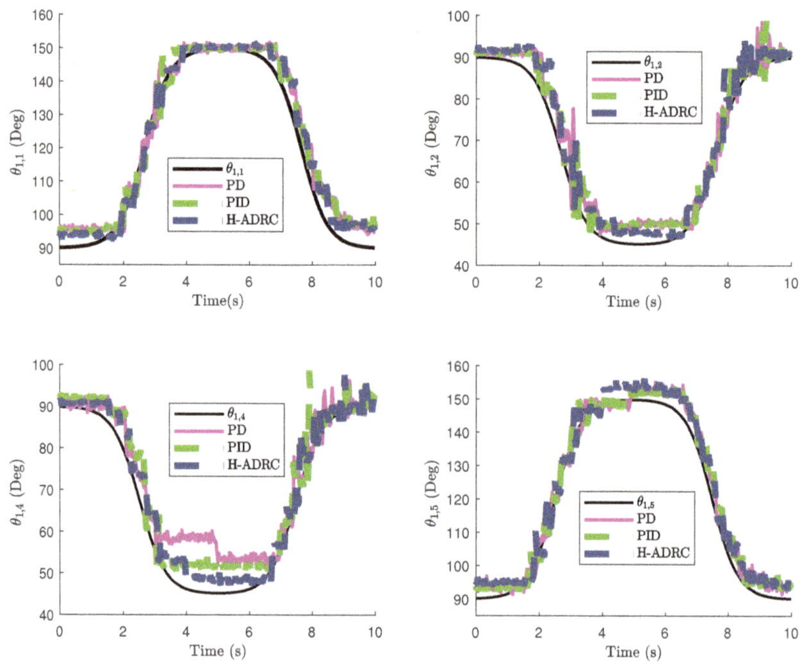

Figure 11. Comparison of the trajectory tracking task in the WBRD by means of different control techniques.

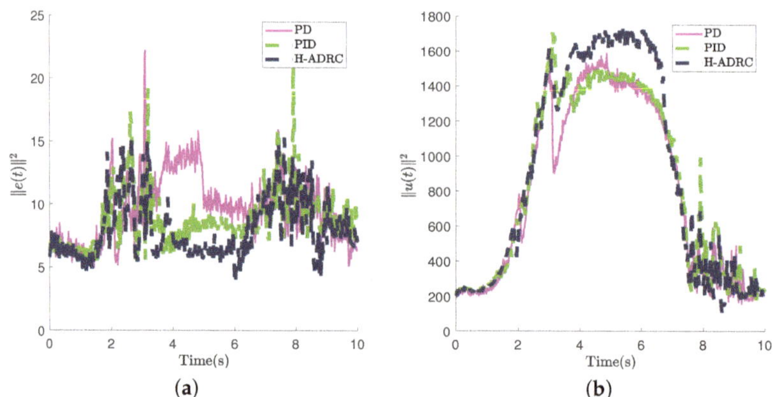

Figure 12. Performance index for each control technique: (**a**) comparison of the Euclidean norm of the tracking error; (**b**) energy used by each controller represented by the evolution of the integral of the square control action u.

Figure 13 provides a sequence of photographs corresponding to the experimental evaluation of the H-ADRC controller over the constructed WBRD. These photos reveal a sequence of the BIMR positions if the controller BIMR regulates the DC motor actuators yielding a coordinated articular movements. The sequence confirms the effects of the additional compensation integrated in the

H-ADRC. Additionally, the sequence confirms the performance of the alternated activation of the vacuum pumps according to the realization of the gait semi-cycle in either scenario a or b.

Experimentally, the hybrid controller provides the efficient tracking of the reference angular positions in both scenarios a and b, despite the model of the WBRD not having been used at all in the experimental sequences.

Figure 14 is intended to highlight the sequence realized by pumps that remain attached to the floor after the proposed controller drives the trajectories toward the references (red arrows for the attached and blue arrows for the released pumps). This strategy succeeded in keeping the angular velocities bounded at each joint in the bioinspired robot. In addition, the reference trajectories for the controller were proposed to keep the distal section of the WBRD closer to the floor. In addition, the absolute values of their time derivatives are small enough to limit the possibility of having fast variations of the controlled angular position at each join, which also contributes to reducing high frequency oscillation of the tracking error. All of these strategies together restricted the possibility of the discontinuous movements' effect on the proposed WBRD.

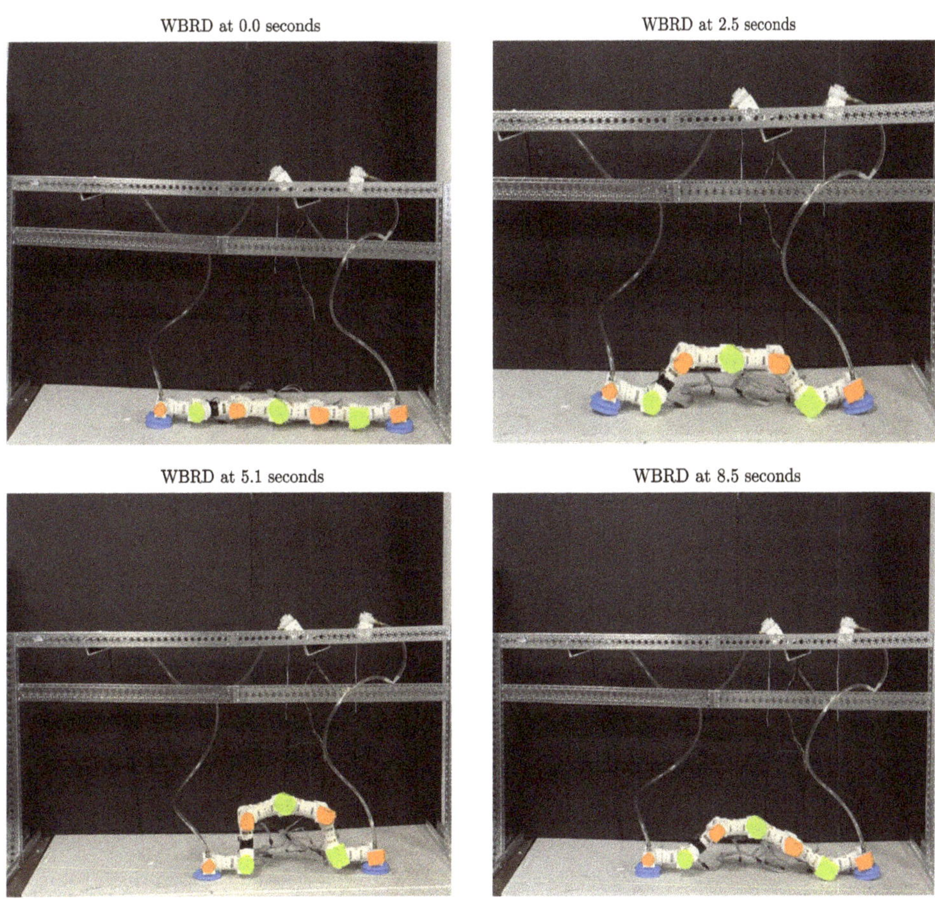

Figure 13. Experimental gait sequence.

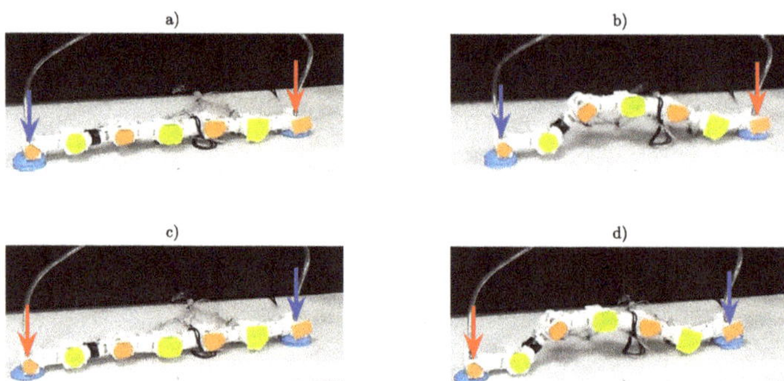

Figure 14. Reference framework switching: (**a**) second vacuum pump activated (red arrow) in zero position; (**b**) second vacuum pump activated during gait cycle; (**c**) first vacuum pump activated in zero position; (**d**) first vacuum pump activated during gait cycle.

Remark 5. *The magnitude of the control signal depends on the initial conditions of the position and velocities of the suggested mobile robot links. Moreover, the magnitude of the control is a trade-off between the convergence time, as well as the accuracy on the tracking error. This problem can be solved by an adaptive version of the controller in order to reduce the control magnitude as the tracking error approaches the origin. In addition, the energy is necessary to fulfill the restrictions imposed by the Barrier function. In the case of the experimental results, the control signal is restricted according to the signal that is sent to the robotic device, which means that a pulse Width modulated (PWM) signal is implemented in the device. The maximum value for the control output is* 2^N *with N the number of bits used for implementing PWM signal. Under this condition, the CD motor is moving as its maximum speed, which is always bounded.*

9. Conclusions

The proposed controller exhibited an acceptable performance even in the presence of parametric uncertainties and noisy measurements. The hybrid structure allows for dealing with the WRBD represented by two link robot manipulators alternating between their first and last links as a reference for its working space. This result constitutes one of the first ADRC approaches dealing at the same time with hybrid systems with restricted variables. A barrier technique imposed angular restrictions in the robotic device avoiding any damage to its physical structure. Moreover, the H-ADRC controller reduces the steady state error compared with classical output feedback structures like state-feedback (PD form) and extended state state feedback (PID structure).

Author Contributions: Conceptualization: I.S., D.C.-O. and I.C.; Methodology: V.L.-O. and A.G.; Software: A.G. and M.M.-S.; Validation: V.L.-O., A.G. and M.M.-S.; Formal analysis, I.S. and D.C.-O.; Investigation: V.L.-O. and D.C.-O.; Resources: I.S. and I.C.; Data curation: V.L.-O. and M.M.-S.; Writing—original draft preparation: V.L.-O. and I.S.; writing—review and editing: I.S. and I.C.; supervision: I.C.; project administration: I.S., D.C.-O. and I.C.; Funding acquisition: I.S., D.C.-O. and I.C. All authors have read and agreed to the published version of the manuscript.

Acknowledgments: The authors acknowledge the support and funding offered by the National Polytechnique Institute via the Grants labeled SIP-2019-5253, SIP-2019-5253 and SIP-2019-6849 .

Conflicts of Interest: Authors declare not any personal interest that may be perceived as inappropriately influencing the representation or interpretation of reported research results.

References

1. Hwang, J.; Jeong, Y.; Park, J.M.; Lee, K.H.; Hong, J.W.; Choi, J. Biomimetics: Forecasting the future of science, engineering, and medicine. *Int. J. Nanomed.* **2015**, *10*, 5701. [CrossRef]
2. Hyun, D.J.; Seok, S.; Lee, J.; Kim, S. High speed trot-running: Implementation of a hierarchical controller using proprioceptive impedance control on the MIT Cheetah. *Int. J. Robot. Res.* **2014**, *33*, 1417–1445. [CrossRef]
3. Wiguna, T.; Heo, S.; Park, H.; Seo Goo, N. Design and Experimental Parameteric Study of a Fish Robot Actuated by Piezoelectric Actuators. *J. Intell. Mater. Syst. Struct.* **2009**, *20*, 751–758. [CrossRef]
4. Kim, S.; Spenko, M.; Trujillo, S.; Heyneman, B.; Santos, D.; Cutkosky, M.R. Smooth Vertical Surface Climbing With Directional Adhesion. *IEEE Trans. Robot.* **2008**, *24*, 65–74. [CrossRef]
5. Seok, S.; Onal, C.D.; Cho, K.J.; Wood, R.J.; Rus, D.; Kim, S. Meshworm: A peristaltic soft robot with antagonistic nickel titanium coil actuators. *IEEE/ASME Trans. Mechatron.* **2013**, *18*, 1485–1497. [CrossRef]
6. Umedachi, T.; Vikas, V.; Trimmer, B.A. Softworms : The design and control of non-pneumatic, 3D-printed, deformable robots. *Bioinspiration Biomim.* **2016**, *11*, 025001. [CrossRef]
7. Qiao, J.; Shang, J.; Goldenberg, A. Development of Inchworm In-Pipe Robot Based on Self-Locking Mechanism. *IEEE/ASME Trans. Mechatron.* **2013**, *18*, 799–806. [CrossRef]
8. Plaut, R.H. Mathematical model of inchworm locomotion. *Int. J. Non-Linear Mech.* **2015**, *76*, 56–63. [CrossRef]
9. Lee, D.; Joe, S.; Choi, J.; Lee, B.I.; Kim, B. An elastic caterpillar-based self-propelled robotic colonoscope with high safety and mobility. *Mechatronics* **2016**, *39*, 54–62. [CrossRef]
10. Jung, G.P.; Koh, J.S.; Cho, K.J. Underactuated adaptive gripper using flexural buckling. *IEEE Trans. Robot.* **2013**, *29*, 1396–1407. [CrossRef]
11. Saab, W.; Kumar, A.; Ben-Tzvi, P. Design and Analysis of a Miniature Modular Inchworm Robot. In Proceedings of the ASME 2016 International Design Engineering Technical Conferences and Computers and Information in Engineering Conference, Charlotte, NC, USA, 21–24 August 2016; pp. V05AT07A060–V05AT07A060.
12. Moreland, S.; Skonieczny, K.; Wettergreen, D.; Asnani, V.; Creager, C.; Oravec, H. Inching locomotion for planetary rover mobility. In Proceedings of the 2011 Aerospace Conference, Big Sky, MT, USA, 5–12 March 2011; pp. 1–6. [CrossRef]
13. Tang, D.; Zhang, W.; Jiang, S.; Shen, Y.; Chen, H. Development of an Inchworm Boring Robot (IBR) for planetary subsurface exploration. In Proceedings of the 2015 IEEE International Conference on Robotics and Biomimetics (ROBIO), Zhuhai, China, 6–9 December 2015; pp. 2109–2114. [CrossRef]
14. Guan, Y.; Jiang, L.; Zhu, H.; Zhou, X.; Cai, C.; Wu, W.; Li, Z.; Zhang, H.; Zhang, X. Climbot: A modular bio-inspired biped climbing robot. In Proceedings of the 2011 IEEE/RSJ International Conference on Intelligent Robots and Systems, San Francisco, CA, USA, 25–30 September 2011; pp. 1473–1478. [CrossRef]
15. Rahmani, M.; Ghanbari, A.; Ettefagh, M.M. Robust adaptive control of a bio-inspired robot manipulator using bat algorithm. *Expert Syst. Appl.* **2016**, *56*, 164–176. [CrossRef]
16. Ayala, H.V.H.; dos Santos Coelho, L. Tuning of PID controller based on a multiobjective genetic algorithm applied to a robotic manipulator. *Expert Syst. Appl.* **2012**, *39*, 8968–8974. [CrossRef]
17. Kolathaya, S.; Ames, A.D. Parameter to state stability of control Lyapunov functions for hybrid system models of 6 robots. *Nonlinear Anal. Hybrid Syst.* **2017**, *25*, 174–191. [CrossRef]
18. Xie, H.; Song, K.; He, Y. A hybrid disturbance rejection control solution for variable valve timing system of gasoline engines. *ISA Trans.* **2014**, *53*, 889–898. [CrossRef] [PubMed]
19. Sira-Ramírez, H.; Luviano-Juárez, A.; Ramírez-Neria, M.; Zurita-Bustamante, E.W. Chapter 2—Generalities of ADRC. In *Active Disturbance Rejection Control of Dynamic Systems*; Sira-Ramírez, H., Luviano-Juárez, A., Ramírez-Neria, M.; Zurita-Bustamante, E.W., Eds.; Butterworth-Heinemann: Oxford, UK, 2017; pp. 13–50. [CrossRef]
20. Tee, K.P.; Ge, S.S.; Tay, E.H. Barrier Lyapunov Functions for the control of output-constrained nonlinear systems. *Automatica* **2009**, *45*, 918–927. [CrossRef]
21. Schaft, A.; Schumacher, H. *An Introduction to Hybrid Dynamical Systems*; Springer: London, UK, 2000.
22. Park, J.H.; Chung, H. Hybrid control for biped robots using impedance control and computed-torque control. In Proceedings of the 1999 IEEE International Conference on Robotics and Automation, Detroit, MI, USA, 10–15 May 1999; Volume 2, pp. 1365–1370.

23. Martin, A.E.; Post, D.C.; Schmiedeler, J.P. Design and experimental implementation of a hybrid zero dynamics-based controller for planar bipeds with curved feet. *Int. J. Robot. Res.* **2014**, *33*, 988–1005. [CrossRef]
24. Cruz, D.; Luviano-Juárez, A.; Chairez, I. Output sliding mode controller to regulate the gait of Gecko-inspired robot. In Proceedings of the Memorias del XVI Congreso Latinoamericano de Control Automático, Cancún Quintana Roo, México, 14–17 October 2014.
25. Bullo, F.; Zefran, M. Modeling and controllability for a class of hybrid mechanical systems. *IEEE Trans. Robot. Autom.* **2002**, *18*, 563–573. [CrossRef]
26. Posa, M.; Tedrake, R. Direct trajectory optimization of rigid body dynamical systems through contact. In *Algorithmic Foundations of Robotics X*; Springer: Berlin, Germany, 2013; pp. 527–542.
27. Merchant, R.; Cruz-Ortiz, D.; Ballesteros-Escamilla, M.; Chairez, I. Integrated wearable and self-carrying active upper limb orthosis. *Proc. Inst. Mech. Eng. H J. Eng. Med.* **2018**, *232*, 172–184. [CrossRef]
28. Sira-Ramírez, H.; López-Uribe, C.; Velasco-Villa, M. Linear Observer-Based Active Disturbance Rejection Control of the Omnidirectional Mobile Robot. *Asian J. Control* **2013**, *15*, 51–63. [CrossRef]
29. Morales, R.; Sira-Ramírez, H.; Feliu, V. Adaptive control based on fast online algebraic identification and GPI control for magnetic levitation systems with time-varying input gain. *Int. J. Control* **2014**, *87*, 1604–1621. [CrossRef]
30. Sira-Ramirez, H.; Nuñez, C.; Visairo, N. Robust Sigma - Delta Generalised Proportional Integral observer based control of a buck converter with uncertain loads. *Int. J. Control* **2010**, *83*, 1631–1640. [CrossRef]
31. Liu, Y.J.; Tong, S. Barrier Lyapunov functions-based adaptive control for a class of nonlinear pure-feedback systems with full state constraints. *Automatica* **2016**, *64*, 70–75. [CrossRef]
32. Yu, X.; Wang, T.; Qiu, J.; Gao, H. Barrier Lyapunov Function-Based Adaptive Fault-Tolerant Control for a Class of Strict-Feedback Stochastic Nonlinear Systems. *IEEE Trans. Cybern.* **2019**. [CrossRef]
33. Poznyak, A. *Advanced Mathematical Tools for Control Engineers: Volume 1: Deterministic Systems*; Elsevier: Amsterdam, The Netherlands, 2010.
34. Salgado, I.; Cruz-Ortiz, D.; Camacho, O.; Chairez, I. Output feedback control of a skid-steered mobile robot based on the super-twisting algorithm. *Control Eng. Pract.* **2017**, *58*, 193–203. [CrossRef]

© 2020 by the authors. Licensee MDPI, Basel, Switzerland. This article is an open access article distributed under the terms and conditions of the Creative Commons Attribution (CC BY) license (http://creativecommons.org/licenses/by/4.0/).

Article

A Fractional High-Gain Nonlinear Observer Design—Application for Rivers Environmental Monitoring Model

Abraham Efraim Rodriguez-Mata [1,*], Yaneth Bustos-Terrones [1], Victor Gonzalez-Huitrón [1], Pablo Antonio Lopéz-Peréz [2], Omar Hernández-González [3] and Leonel Ernesto Amabilis-Sosa [1]

1 Conacyt-Tecnológico Nacional de México\Instituto Tecnológico de Culiacán, Juan de Dios Bátiz 310, Col. Guadalupe, Culiacán Rosales CP 80220, Sinaloa, Mexico; yabustoste@conacyt.mx (Y.B.-T.); victor.gonzalez@itculiacan.edu.mx (V.G.-H.); lamabilis@conacyt.mx (L.E.A.-S.)

2 Escuela Superior de Apan, Universidad Autonoma de Hidalgo, Carretera Apan-Calpulalpan s/n, Colonia, Chimalpa Tlalayote 43920, Hidalgo, Mexico; save1991@yahoo.com.mx

3 Conacyt-Tecnológico Nacional de México\Instituto Tecnológico de Hermosillo, Av. Tecnológico y, Periférico Poniente S/N, Sahuaro, Hermosillo 83170, Sonora, Mexico; ohernandezg@conacyt.mx

* Correspondence: aerodriguez@conacyt.mx

Received: 2 June 2020; Accepted: 10 July 2020; Published: 16 July 2020

Abstract: The deterioration of current environmental water sources has led to the need to find ways to monitor water quality conditions. In this paper, we propose the use of Streeter–Phelps contaminant distribution models and state estimation techniques (observer) to be able to estimate variables that are very difficult to measure in rivers with online sensors, such as Biochemical Oxygen Demand (BOD). We propose the design of a novel Fractional Order High Gain Observer (FOHO) and consider the use of Lyapunov convergence functions to demonstrate stability, as it is compared to classical extended Luenberger Observer published in the literature, to study the convergence in BOD estimation in rivers. The proposed methodology was used to estimated Dissolved oxygen (DO) and BOD monitoring of River Culiacan, Sinaloa, Mexico. The use of fractional order in high-gain observers has a very effective effect on BOD estimation performance, as shown by our numerical studies. The theoretical results have shown that robust observer design can help solve problems in estimating complex variables.

Keywords: Streeter–Phelps model; fractional-order control; high observers; river monitoring

1. Introduction

Among the main problems of humanity are those related to water availability and pollution. Unfortunately, industrial development and population growth lead to an increase in pollutant discharges which have a negative impact mostly on aquatic ecosystems [1]. With the accelerated urbanization in the world, and especially in Mexico, environmental water problems have become even more considerable. Long-term ineffective governance has driven to aggravate water pollution in some areas more than ever before [2]. In recent years, with importance on environmental water quality, water pollution has been gradually brought under control. In addition, research has been centred on the field of water quality exploration and pollutant diffusion simulation in rivers and watersheds via mathematical modeling and novel tech [3].

Related to aquatic environments researches, the water quality model is a fundamental strategy for the water study and its forecasts. Streeter–Phelps model is distinguished as the first dynamic and spatial water quality model. Therefore, almost all modern timeline researchers have done an enormous amount of research in improving and developing water quality models [4]. Due to their relevance, experts in environmental river control are still employing it to obtain mathematical models of water quality from rivers and lakes [5–9].

The Streeter–Phelps model relates the two main mechanisms that define dissolved oxygen in a lotic water Biochemical Oxygen Demand (BOD) receiving wastewater tributary. These mechanisms are a decomposition of organic matter and oxygen aeration [5]. It is also used to estimate BOD and Dissolved oxygen (DO) transport, which is achieved by feeding data on hydromorphological and water quality parameters [10].

Usually, automated water quality control in rivers and lakes is still done traditionally, i.e., employing complicated laboratory techniques, which implies time, consumable resources, and delays in decision-making [11]. Some robotic prototypes have been proposed to obtain a better environmental control [12–14]; these are an excellent tool for active monitoring of water quality since this type of robotic device has embedded sensors of physicochemical parameters of remote-controlled GPS position [15–17]. Currently, existing commercial sensors can measure variables, such as turbidity, pH, temperature, and DO, which help measure water quality in water bodies [14]. Currently, there are remarkable scientific advances in relation to mobile and fixed stations for environmental monitoring of rivers [18–20]. Still, there are water quality variables that are not commonly measured by commercial sensors, such as chemical oxygen demand and BOD [11]. Using the information about the *Streeter-Phelps* model and algorithms called Observers, it is possible to estimate BOD via the information of other sensors, such as the available DO and longitude (GPS signal). Observers have received several versions and improvements in their performance, over time, in the difficult task of state estimation and parameters in complex systems, but these are still being improved in all their different applications. For example, some outstanding works of each technique, from the first generations linear [21], non-linear [22], adaptive [23], sliding mode [24], non-uniformly observable [25], and discreet observers who can provide good results [26], based on the frequency [27] and high-gain systems [28], have been proposed. There are also applications of its functions and fuzzy techniques for diagnosing sensor faults and multiple applications [29,30]. Currently, the use of fractional techniques and their implementation in state estimation via observers is being arduously studied, since the use of fractional techniques allows to add degrees of freedom in the tuning of the observers [31].

From the knowledge of the authors, there are few published works about the use of observers for *Streeter-Phelps* states estimation; there is a very outstanding work where it was used as a linear technique and a Luenberger observer for the BOD estimation [32]. This remarkable work was possible due to change in the basis to obtain an observable model. However, given the time in which it was published, obsolete techniques are used that are not robust to parametric changes and in the presence of disturbances.

In this paper, we present a robust Fractional High-Gain Nonlinear Observer design for state estimation in the Streeter–Phelps model published in Reference [32]. Besides, we analyze parametric sensibility via the use of Lyapunov convergence functions. The proposed novel fractional observer design is based on a Lyapunov analysis. The proposal is directed to future work where this kind of algorithms can be embedded and programmed in robotic watercraft to estimate environmental rivers quality water variables. MATLAB Simulink simulations are made to show the advantages of our algorithm over those proposed for similar models. Finally, in conclusion, there are some perspectives and thoughts open to investigation for possible tasks in real time.

2. Mathematical Model and Problem Statement

The most common nutrient and oxygen distribution pollutant mathematical model is the *Streeter-Phelps* system. Since 1925, it has described the oxygen balance in rivers, lakes, and water sources. Although the appearance of more complex models, the *Streeter-Phelps* model (and its extensions) continues to be one of the most widely used, since it is a relatively simple mathematical model. However, it injects much essential information; despite almost a century since its presentation, it continues to be used [4–10]: Under some assumptions, *Streeter-Phelps* model helps to estimate and measure the main variables for environmental studies, such as BOD and DO, via aquatic robots and algorithms observers programmed, thus improving the environmental monitoring of rivers. Together, BOD and DO are the parameters that represent the largest number of pollutants in water bodies. Indeed, the higher the BOD, the lower the amount of OD, which in turn poses a greater risk to aquatic life. Furthermore, the study of these parameters allows us to know the resilience of a river and/or the degree of pollution received by polluting discharges [33]. We propose to use the model in this work under the following assumptions.

Condition 1.

1. *The river maintains a constant volumetric flow rate and possible bounded changes to nominal flow rate (U volume/time).*
2. *It is assumed that a one-dimensional longitudinal displacement in the same natural direction as the river direction of the river (L longitude). From $L_0 = 0$ to $L \to \infty$.*
3. *It senses dissolved oxygen ($x_1(L) = x_1$) $mgO_2/volume$) in real-time in an embedded manner.*
4. *BOD ($x_2(L) = x_2$) $mgO_2/volume$) can be estimated via state or observer estimation techniques, as the $y = x_1$ output is available.*
5. *Based on the literature [34], we will assume that the mathematical models parameters are invariant to time but sensitive to changes in position $L > 0$ or other factor as: temperature, flow velocity, etc., i.e., in this paper, we propose L as the independent variable.*

Therefore, given the nature of phenomenology, the independent variable throughout this work will be the position L. We have the following Streeter–Phelps mathematical model respect position [35]:

$$\begin{aligned} \frac{dx_1}{dL} &= -\frac{k_1(L)}{U}x_2 + \frac{k_2(L)}{U}(D_s - x_1)) \\ \frac{dx_2}{dL} &= -\frac{k_1(L)}{U}x_2, \end{aligned} \tag{1}$$

where the parameters $k_1(L) > 0$ are the BOD removal rate coefficient, $k_2(L) > 0$ is the reaeration coefficient, and D_s is the oxygen saturation level. This type of system tends to be extremely sensitive to parameter changes, in this case, parameters, such as linear velocity U and the variation of constants $k_{1,2}$ [33]. Therefore, it is necessary to perform a parametric sensitivity analysis, which will be substantial to design a state observer. In this paper, we propose such analysis by using stability concepts in the sense of Lyapunov. For the above, it is necessary to make some conceptual assumptions to study stability analysis.

Condition 2. *It is proposed that the model parameters (1) fulfilled following condition:*

$$\begin{aligned} k_1(L) &= k_1 + \beta_1(L) \\ k_2(L) &= k_2 + \beta_2(L), \end{aligned} \tag{2}$$

where $k_{1,2}$ are nominal known parameters; $\beta_1(L)$ and $\beta_2(L)$ are continuously bounded functions $|\beta_1(L)| < \alpha_1 l$ and $|\beta_2(L)| < \alpha_2 L$; and the $\alpha_{1,2} > 0$ are Lipschitz constants respect to longitude L. In this paper, we propose that $\beta_1(L), \beta_2(L)$ functions are unknown, unexpected but bounded, which cause unwanted parametric distortion $\delta(L)$.

Substituting (2) on the mathematical model (1) and reducing:

$$\begin{aligned}
\frac{dx_1}{dL} &= -\frac{k_1}{U}x_2 - \frac{k_2}{U}x_1 + \delta_1 + \frac{k_2}{U}D_s \\
\frac{dx_2}{dL} &= -\frac{k_1}{U}x_2 + \delta_2 \\
\delta_1 &= -\frac{1}{U}(-\beta_1(L)x_1 - \beta_2(l)(D_s - x_2)) \\
\delta_2 &= -\frac{1}{U}\beta_1(L)x_1.
\end{aligned} \tag{3}$$

The previous system (3) can be represented in its state form as a autonomous linear system in bounded disturbance presence:

$$\frac{dx}{dL} = Ax + \delta(L) \tag{4}$$

for $x^T = [x_{1,}\, x_2]^T$, $\delta^T(L) = \frac{1}{U}[\delta_1 + k_2 D_s, \delta_2]^T$, and $A = \begin{bmatrix} -k_2\frac{1}{U} & -k_1\frac{1}{U} \\ 0 & -k_2\frac{1}{U} \end{bmatrix}$. Hence, a general condition to the global disturbance $\delta(L)$ is proposed.

Condition 3. *Let $\delta(L)$ be a bounded unknown nonlinear function such that the following is fulfilled:*

$$\|\delta(L)\| \leq D_{\max} > 0. \tag{5}$$

Note 1. *In this paper, we will assume that the linear velocity remains constant, as we focus on the possible change in kinetic, chemical parameters ($k_{1,2}$). The parametric distortions due to linear velocity (U_{river}) changes will not be treated in this work due to an extension of it. But, it is easy to see that this kind of distortions (linear velocity changes) would cause limited delta distortions; therefore, for $U_{river} = U_{nom} + \delta(U)U$:*

$$\begin{aligned}
\frac{dx}{dL} &= A_{U_{nom}}x + \delta(U)X \\
\|\delta(U)\| &\leq \|A_{U_{nom}}\|.
\end{aligned} \tag{6}$$

Therefore, the following theoretical result is proposed, where parametric sensitivity is analyzed via a stability analysis using Lyapunov's convergence functions.

Theorem 1. *Let $x = 0$ be the spatial equilibrium point of (4) and $\delta(L)$ fulfilled above Condition 3. The origin $x = 0$ will maintain a bonded dynamic in a ball of convergence in parametric chances presence for non-negative matrix Q and P such that: $\lambda_{\min}\{Q\} >> 2D_{\max}\lambda_{\max}\{P\}$.*

Proof. The following Lyapunov function is proposed: $V = \frac{1}{u}x^T Px$ with independent variable L, if it replaces the derivatives under trajectories systems:

$$\begin{aligned}
\frac{dV}{dL} &= \frac{1}{U}\dot{x}^T Px + \frac{1}{U}x^T P\dot{x} \\
\frac{dV}{dL} &= x^T(PA + A^T P)x + 2x^T P\delta(t).
\end{aligned} \tag{7}$$

Since A of (4) is Hurwitz matrix for $U, k_1, k_2 > 0$, then it is $\mathrm{Re}\,\lambda_i > 0$ (due to the nature non-negative parametric system). Hence, Lyapunov algebraic equation calculation $PA + A^T P = -Q$ is fulfilled.

Therefore, (7) is reduced, with $c = 2D_{\max}\lambda_{\max}\{P\}$, $\lambda_{\min}\{Q\} = a$ and bounded:

$$\frac{dV}{dL} = -x^T Q x + 2x^T P \delta(t)$$
$$\frac{dV}{dL} \leq -a\left\|x\right\|^2 + c\left\|x\right\| \qquad (8)$$
$$\frac{dV}{dL} \leq -a\left\|x\right\|\left(\left\|x\right\| - \frac{c}{a}\right).$$

The higher the value of the system, the faster it will converge in the neighborhood to a ball of convergence such that

$$B_x = \{\|x\| \in \mathbb{R}^2 : \|x\| < \frac{c}{a}\} \qquad (9)$$
$$B_x = \{\|x\| \in \mathbb{R}^2 : \|x\| < \frac{2D_{\max}\lambda_{\max}\{P\}}{\lambda_{\min}\{Q\}}\}.$$

In the disturbances system presence, (4) in spatial equilibrium point will remain Ultimated Bounded [36]. □

In order to exemplify the previous theorem use, as well as P, Q matrix calculation, we will use real data obtained from literature and be able to make a parametric stability analysis and sensitivity in the theorem above sense. $U = 1$ is proposed, and it is assumed that the parameters of k_1, k_2 have a 5% maximum change from the nominal value.

First, via Condition 2, the maximum bounded is obtained $|\beta_1(l)| = 0.05k_{1,nom}$ and $|\beta_2(l)| = 0.05k_{2,nom}$. Thus, by replacing this in (4) with nominal parameters, $k_1 = 0.3$ day^{-1} and $k_2 = 0.06$ day^{-1} of literature [34]. Disturbance vector $\delta(L)$ will be calculated:

$$\delta(L) = \begin{pmatrix} \beta_1(l)x_1 \\ \beta_1(l)x_1 - \beta_2(t)x_2 \end{pmatrix}$$
$$\delta(L) = \begin{pmatrix} -0.03x_1 \\ 0.03x_1 - 0.006x_2 \end{pmatrix}. \qquad (10)$$

If parameters are substituted, in matrix A (4), it is obtained:

$$A = \begin{bmatrix} -0.3 & 0 \\ -0.3 & -0.06 \end{bmatrix}.$$

Therefore, there is a $\Lambda = 0.05A$ matrix, such that $\delta(L) = \Lambda x$, such that:

$$\delta(L) = \underbrace{\begin{bmatrix} -0.015 & 0 \\ 0.015 & -0.003 \end{bmatrix}}_{\Lambda.} x$$

If the matrix $Q = -I$ is chosen, Lyapunov P matrix calculation is obtained by matrix system solution $PA + A^T P = -I$, such that:

$$P = \begin{bmatrix} 1.6667 & -1.3889 \\ -1.3889 & 15.2778 \end{bmatrix}, \qquad (11)$$

where the eigenvalues of P are $\mathrm{Re}\,\lambda_i = [1.667 \quad 90.2778]$, to obtain the convergence ball of (9), in the presence of a constant perturbation of 5%, such that $\lambda_{\max}\{P\} = 15.2778$, $\lambda_{\max}\{Q\} = 1$ and $\lambda_{\max}\{\Lambda\} = 0.03$. In addition:

$$\begin{aligned} \|\delta(L)\| &= \|\Lambda x\| \\ \|\delta(L)\| &\leq 0.05\,\|A\|\,\|x\| \leq 0.05\lambda_{\max}\{\Lambda\}\,\|x\|\,0.015\,\|x\| \\ D_{\max} &\leq 0.015\,\|x\|\,. \end{aligned} \tag{12}$$

Substituting (12) on the inequality calculation (8):

$$\begin{aligned} \frac{dV}{dL} &\leq -a\,\|x\|^2 + c\,\|x\| \leq -\lambda_{\min}\{Q\}\,\|x\|^2 + 2D_{\max}\lambda_{\max}\{P\}\,\|x\| \\ \frac{dV}{dL} &\leq -\|x\|^2 + 0.4581\,\|x\|^2 \leq -0.5419\,\|x\|^2\,. \end{aligned}$$

Since Lyapunov's function $V = \frac{1}{u}x^T Ax$ and Raleigh's inequality $\frac{u}{\lambda_{\max}\{A\}}V \leq \|x\|^2$, it is had:

$$\frac{dV}{dL} \leq -1.80V.$$

Therefore, finally:

$$V = V(0)e^{-1.80l}.$$

For a constant perturbation, the system reaches *asymptotic stability*, which is a sub-case of *practical stability*. When perturbation varies from the position, the system will converge to a ball of convergence in the sense of Reference [36].

Problem Statement

The system represented in the linear system (4) has analytical properties that do not allow state estimators to be used, since the y output is a signal that does not permit all states reconstruction, due to general observability matrix, has not been full range, and this has already been studied in the literature. Through certain analytical assumptions about the system (4), a non-linear observable system is obtained, with which it is possible to estimate variables via the use of an observer [32]. By assuming a linear relationship between the BOD removal rate and DO, with zero DO giving no decrease in BOD, a new removal rate coefficient is defined by $k_3 = \frac{k_1}{x_1}$. These modifications, performed in (1), show:

$$\begin{aligned} U\frac{dx_1}{dL} &= -k_1x_2 + k_2(D_s - x_1) + I_D - c_1x_1 \\ U\frac{dx_2}{dL} &= -k_3x_1x_2 + I_B - c_1x_2, \end{aligned} \tag{13}$$

where I_D is the inflow of DO, I_B the BOD inflow, c_1 the oxygen turnover rate due to flow through, and, on the average, $I_D = c_1x_1$ and $I_B = c_1x_2$. Thus, the previous system can be rewritten as follows:

$$\begin{aligned} \frac{dX}{dL} &= A_0X + f(X) \\ y &= x_1, \end{aligned} \tag{14}$$

where $A_0 = \begin{pmatrix} 0 & 1 \\ 0 & 0 \end{pmatrix}$, and $f(X) = \begin{pmatrix} -x_2 + \frac{1}{U}(-k_1x_2 + k_2(D_s - x_1) + I_D - c_1x_1) \\ \frac{1}{U}(-k_3x_1x_2 + I_B - c_1x_2) \end{pmatrix}$.

For robust BOD estimation via the above model, we propose a novel fractional observer, and its analytical design is proposed in the following section.

3. Design of a Novel Fractional Observer (FOHO)

Firstly, fractional calculus is the expansion of traditional analysis to derivation and integration operations employing non-integer orders. In this paper, a fractional-order high gain observer (FOHO) is proposed in order to estimate non-measurable variables (BOD) of a river basin environmental monitoring problem. The Caputo fractional derivative of order β of a function $\phi(L)$ on the positive real half axis is determined. It is important to notice that the notation $D^{-\beta}\phi(L)$ is used, as well as to denote the fractional integral of order β of function $\phi(L)$, more precisely $D^{-\beta}\phi(L) \equiv I^{\beta}\phi(L)$ [37]. The definitions of fractional derivative and fractional integral, as stated above, cannot be used in practice; thus, numeric methods, such as the one based on the Grünwald–Letnikov approach, are commonly used. Here, we deal with the same note in terms of the definition of the fragmentary calculation tool as in Reference [38]. Therefore:

$$I^{\beta}\phi(L) = \frac{1}{\Gamma(\beta)} \int_0^L \frac{\phi(\tau)}{(L-\tau)^{1-\beta}} d\tau, \tag{15}$$

where $0 < \tau < L$ and β. Thus, the following fractional observer (FOHO) is proposed for system (14):

$$\begin{array}{rl} \frac{dz}{dL} = & A_o z + f(z) + bu + \Lambda(\theta) C^T C e + \hat{\delta} \\ \hat{\delta}\,(L) = & -S\Delta_{\theta}^{-1} I^{\beta} sign(e) \end{array} . \tag{16}$$

For $e = Z - X$, it is as follows:

$A_o \in \mathbb{R}^{n \times n}$	is the state matrix in canonical form,
$z(t) \in \mathbb{R}^{n \times 1}$	is the estimate state vector,
$\hat{\delta}(L) \in \mathbb{R}^{n \times 1}$	is the estimate state disturbance,
$\Lambda(\theta) \in \mathbb{R}^{n \times n}$	is the high gain Hurwitz matrix,
$\beta \in \mathbb{R}$	is the fractional order integer,
$\Delta_\theta \in \mathbb{R}^{n \times n}$	is the symetric matrix such that $\Delta_\theta = diag(1, \theta, \theta^2 \ldots \theta^{n-1})$, and
$S \in \mathbb{R}$	is a Lyapunov symmetric matrix,

where $h_1, ..., h_n > 0$ are gains. h_i is an observer gain, θ is a positive high gain, and $\Lambda(\theta)$ matrix is as follows:

$$\Lambda(\theta) = \begin{bmatrix} \theta h_1 & 1 & 0 & \cdots & 0 \\ \theta^2 h_2 & 0 & 1 & \ddots & \cdots \\ \vdots & \vdots & \ddots & \ddots & \vdots \\ \theta^{n-1} h_{n-1} & 0 & \cdots & 0 & 1 \\ \theta^n h_n & 0 & \cdots & 0 & 0 \end{bmatrix}.$$

Hence, there is a matrix A_h such that it is fulfilled:

$$\Delta_\theta^{-1} \Lambda(\theta) \Delta_\theta = \theta A_h. \tag{17}$$

There is a symmetric Δ_θ such that $\Delta_\theta = diag(1, \theta, \theta^2 \ldots \theta^{n-1})$. Thus, A_h is a Hurwitz matrix by construction, such that:

$$A_h = \begin{pmatrix} h_1 & 1 & 0 & \cdots & 0 \\ h_2 & 0 & 1 & \ddots & \vdots \\ \vdots & \vdots & \ddots & \ddots & 0 \\ h_{n-1} & 0 & 0 & 0 & 1 \\ h_n & 0 & 0 & 0 & 0 \end{pmatrix}. \tag{18}$$

Condition 4. *Let $e = Z - X$ be estimated error such that the following is fulfilled:*

1. *A_h will be a Hurwitz matrix always exists a matrix $S = S^T$ such that [39]:*

$$SA_h + A_h^T S = -I. \tag{19}$$

Therefore, for any $\theta > 1$, introduce the next symmetric matrix:

$$S_\theta = \Delta_\theta^{-1} S \Delta_\theta^{-1}.$$

2. *Consider the error function $e = z - x$ of (16) and (14). Existing $\lambda_{\min}\{S\} = c > 0$ such that [40]:*

$$c\left|I^\beta sign(e)\right| > \delta_{\max}. \tag{20}$$

Therefore, the main general theoretical result of this work is presented by means of the following theorem.

Theorem 2. *Convergence of FOHO estimator. Let an observer (16) and (14) be, it is said, that general error of state estimation $e(t) = z(L) - x(L)$ converge asymptomatically, with high gain function $\theta > 1$ in unknown nonlinear perturbation presence $|\delta(t)| < D_{\max}$ if A_h is a Hurwitz matrix with $D_{\max}$ known, and matrix $C = [1\,0\,0\ldots0]$ if it is fulfilled Condition 4.*

Proof. The estimate error $\frac{de}{dL} = \frac{dz}{dL} - \frac{dx}{dL}$ has been substituted in the dynamics Equations (16) and (14) corresponding:

$$\begin{aligned}
\frac{de}{dL} &= A_0 z + bu + \hat{\delta} + \Lambda(\theta)C^T Ce - A_0 x - bu - \delta + k(\cdot)\\
\frac{de}{dL} &= (A_0 + \Lambda(\theta)C^T C)e + \tilde{\delta} + k(\cdot)\\
\frac{de}{dL} &= (A_0 + \Lambda(\theta)C^T C)e + \tilde{\delta} + k(\cdot),
\end{aligned} \tag{21}$$

where $k(\cdot) = f(Z) - f(X)$ via Lipschitz relationship, and the following condition can be obtained [39], such that: $\|k(\cdot)\| \le \gamma(z - x) \le \gamma e$. In addition, it has the following remarkable matrix properties:

$$\Lambda(\theta)C^T C = \begin{bmatrix} \theta h_1 & 1 & 0 & \cdots & 0\\ \theta^2 h_2 & 0 & 1 & \ddots & \cdots\\ \vdots & \vdots & \ddots & \ddots & \vdots\\ \theta^{n-1}h_{n-1} & 0 & \cdots & 0 & 1\\ \theta^n h_n & 0 & \cdots & 0 & 0\end{bmatrix}\begin{bmatrix} 1 & 0 & 0 & \cdots & 0\\ 0 & 0 & 0 & \ddots & \cdots\\ \vdots & \vdots & \ddots & \ddots & \vdots\\ 0 & 0 & \cdots & 0 & 0\\ 0 & 0 & \cdots & 0 & 0\end{bmatrix} = \begin{bmatrix} \theta h_1 & 0 & 0 & \cdots & 0\\ \theta^2 h_2 & 0 & 0 & \ddots & \cdots\\ \vdots & \vdots & \ddots & \ddots & \vdots\\ \theta^{n-1}h_{n-1} & 0 & \cdots & 0 & 0\\ \theta^n h_n & 0 & \cdots & 0 & 0\end{bmatrix}.$$

Therefore, $A_0 + \Lambda(\theta)C^T C = \begin{bmatrix} \theta h_1 & 1 & 0 & \cdots & 0\\ \theta^2 h_2 & 0 & 1 & \ddots & \cdots\\ \vdots & \vdots & \ddots & \ddots & \vdots\\ \theta^{n-1}h_{n-1} & 0 & \cdots & 0 & 1\\ \theta^n h_n & 0 & \cdots & 0 & 0\end{bmatrix} = \Lambda(\theta).$

Hence, (21) is modified:

$$\frac{de}{dL} = \Lambda(\theta)e + \tilde{\delta} + k(\cdot). \tag{22}$$

It is proposed a Lyapunov function $V_\theta(e) = V_\theta$ with its derivates along trajectories of (22):

$$\begin{aligned}
V &= e^T S_\theta e\\
\frac{dV}{dL} &= e^T(S_\theta \Lambda(\theta) + \Lambda^T(\theta)S_\theta)e + 2e^T S_\theta \tilde{\delta} + 2e^T S_\theta k(\cdot).
\end{aligned} \tag{23}$$

Recalling Condition 4 $S_\theta = \Delta_\theta^{-1} S \Delta_\theta^{-1}, e_\theta = \Delta_\theta^{-1} e$, and $e_\theta^T = (\Delta_\theta^{-1} e)^T$ and reshuffling terms:

$$\frac{dV}{dL} = e_\theta^T[S\Delta_\theta^{-1}\Lambda(\theta)\Delta_\theta + (\Delta_\theta^{-1}\Lambda(\theta)\Delta_\theta)^T S]e_\theta + 2e_\theta^T Sk(\cdot) + 2e_\theta^T S\Delta_\theta^{-1}\tilde{\delta}. \quad (24)$$

Replacing (17) on (24):

$$\frac{dV}{dL} = \theta e_\theta^T[SA_h + A_h^T S]e_\theta + 2e_\theta^T Sk(\cdot) - 2e_\theta^T S\Delta_\theta^{-1}\tilde{\delta}. \quad (25)$$

Giving $\tilde{\delta} = \hat{\delta} - \delta$, if we substitute the $\hat{\delta}$ of observer (16):

$$\frac{dV}{dL} = \theta e_\theta^T[SA_h + A_h^T S]e_\theta + 2e_\theta^T Sk(\cdot) - 2e_\theta^T S\Delta_\theta^{-1}(2S\Delta_\theta^{-1}I^\beta sign(e) - \delta). \quad (26)$$

Given that A_h is a Hurwitz matrix and $\|\delta\| \leq D_{\max}$ and fulfilling Condition 4, it has, as follows:

$$\begin{aligned}
\frac{dV}{dL} &= -\theta e_\theta^T e_\theta + 2e_\theta^T Sk(\cdot) - 2e_\theta^T S\Delta_\theta^{-1}(2S\Delta_\theta^{-1}I^\beta sign(e) - \delta) \\
\frac{dV}{dL} &= -\theta e_\theta^T e_\theta + 2e_\theta^T Sk(\cdot) - e_\theta^T S\Delta_\theta^{-1}(S\Delta_\theta^{-1}I^\beta sign(e) - \delta) \\
\frac{dV}{dL} &\leq -(\theta - b)\,\|e\|_\theta^2 + c\,\|e\|_\theta\,(c\left\|I^\beta sign(e)\right\| - D_{\max}).
\end{aligned} \quad (27)$$

Since $\lambda_{\max}\{\Delta_\theta^{-1}\} = 1$ for one high gain $\theta \gg b$, where $b = 2\lambda_{\min}\{S\}\lambda_{\min}, c = \lambda_{\max}\{S\}\lambda_{\max}\{\Delta_\theta^{-1}\} = \lambda_{\min}\{S\}$, the observer will converge asymptomatically to zero when the longitude tends to infinity, in same sense as Reference [37,40]. □

4. Materials and Methods

In order to carry out the numerical simulation study, it is necessary to mention the conditions and system parameters to be studied, in this case, an urban river in Mexico. All these data were obtained from the literature; thus, it is possible to carry out a theoretical numerical simulation to be able to appreciate the β parameter observer effect on the BOD estimate in the river simulation. The Culiacán river, located in the State of Sinaloa, is of great regional importance since this country region is known as the granary of Mexico, and this river feeds the agricultural industry. Its course is 87.5 km long, and its basin covers 17,200 km^2, with an annual flow of 3280 million hm^3. It has an average depth of 1335 m, with an average width of 50 m and a flow of 0.5 m^3/s [41,42].

The river is formed at the confluence Tamazula and Humaya rivers. It runs along the Pacific coastal plain, initially flowing through a large part of the urban area, which is why it has been indicated that the Culiacán River has a certain level of pollution resulting from the discharge of contaminated water from industrial processes [43]. Therefore, it is necessary to be able to monitor water quality and its DO and BOD levels. In this section, we suggest a numerical simulation study where we suggested estimation BOD and monitoring of DO in presence to Streeter-Phelps model parameters changes. The modelling parameters are taken from those published in Reference [32], and we will have 100% k_1 and k_2 nominal parameters along the river position L, simulating the course of an aquatic boat robot at constant velocity in the environmental monitoring task by measuring DO (x_1) and estimating BOD (x_2) using the fractional observer shown (16).

In this paper, we performed a simulation study on the use of the Streeter–Phelps model for DO measurement and BOD estimation of the Culiacán river. The simulation was carried out in MATLAB-Simulink language, via a numerical method ODE45 with variable step. Based on the physical data model river parameters (14), the study was carried out using the simulation parameters summarised in the following Table 1.

Table 1. Model (14) and observer (16) simulation parameters.

Parameter	Value
Nominal k_1	9.6 L/mo
Nominal k_2	6 L/mo
U	7 km/mo
I_D	347 mg/L*mo
I_B	197.8 mg/L*mo
D_s	15.76 mg/L
$x_1(0)$	10 mg/L
$x_2(0)$	3.83 mg/L
$z_1(0)$	10 mg/L
$z_2(0)$	4.2 mg/L
h_1	-1.5
h_2	-5
θ	2
β	[0, 0.3, 0.6, 0.9, 1]
Ł	$0 \rightarrow 2$ km

To test FOHO robustness (16) and Luenberger classical [32] observer algorithms, we noted a parameter change throughout the simulation, recreating changes along river length (see to Figure 1). We suggest changes on the Culiacán nominal Streeter–Phelps parameters from the position or length of the river. It is proposed that the DO signal (x_1) recreates more realistic conditions; therefore, we add a white Gaussian noise signal. Since a theoretical numerical study is presented in this work, the $L = 0$ is arbitrary, and the initial conditions of BOD and DO are based on the state-of-the-art. Similarly, it is assumed that there are no additional pollution sources within the 2 km proposed in this numerical simulation.

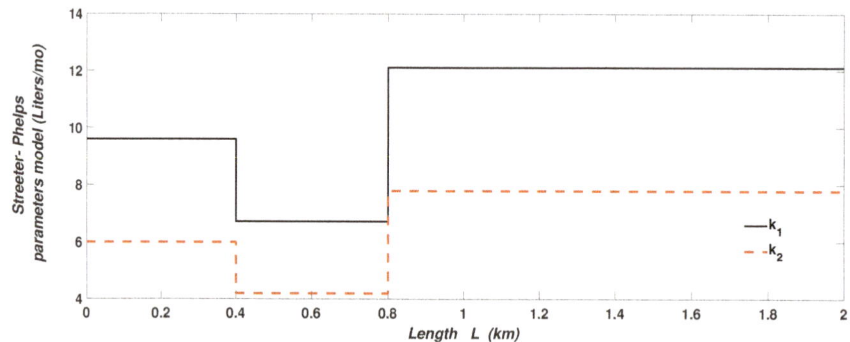

Figure 1. Proposed changes in parameters.

5. Results

In this brief chapter, we show the numerical simulation results carried out. We study the effect of the fractional observer order in presence of parameters changed. Simulating the spatial change dynamics in DO and BOD concentrations (see Figures 2 and 3), through that of the FOHO and the classical Luenberger observer [32], we can appreciate β parameter change of FOHO; this is unperceivable at a glance. Still, we can significantly understand the FOHO obtains better performance over its classical counterpart. The FOHO (16) rejects Streeter–Phelps parameter changes with significant performance.

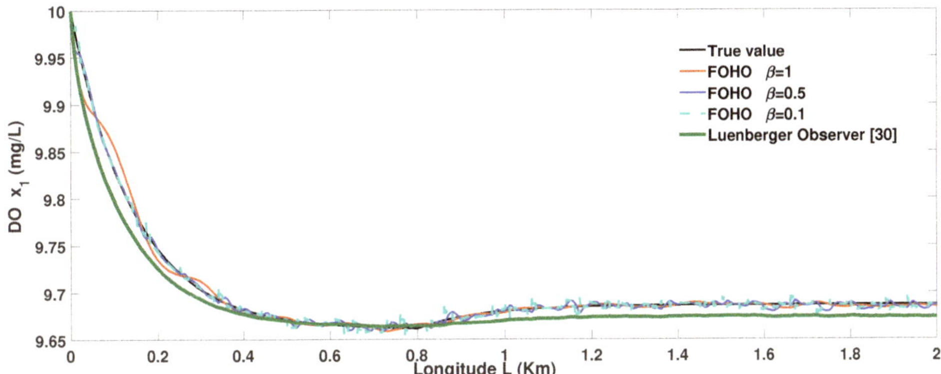

Figure 2. Comparison of Dissolved oxygen (DO) monitoring estimate: the fractional observer (16) and the extended Luenberger classical observer.

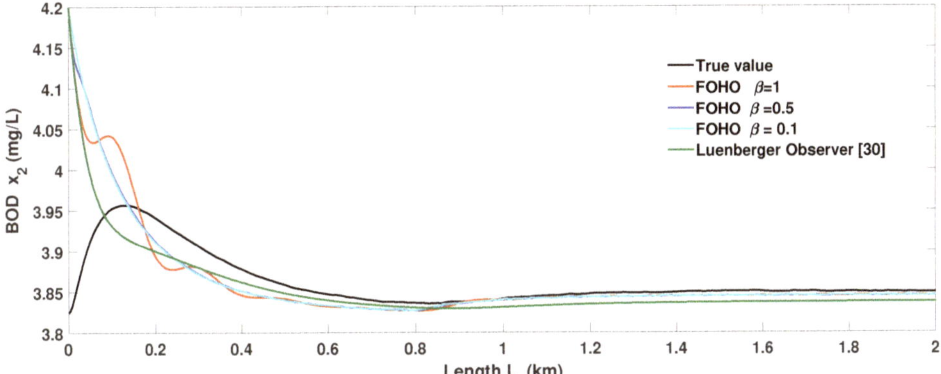

Figure 3. Biochemical Oxygen Demand (BOD) monitoring using observer (16) and its changes depending on β fractional-order value and classic observer.

To quantify the performance and accuracy of the state observers, criteria, such as the Integral Distance Absolute error (IDAE), can be used to evaluate the estimation of states in processes of chemical or biochemical character usually used in the evaluation of observers [44–46]. To evaluate the effect of the β performance and accuracy of the state observers, we use two different types of error criteria, in same sense as Reference [44].

1. Steady-state performance factor (SSPF) ranges from 0 to 1, where zero indicates perfect process state reconstruction. Perfect reconstruction in this context means that the process variables in the estimated state vector coincide with the true process. The steady-state performance factor is defined as:

$$SSPF = \frac{|x(t) - \hat{x}(t)|}{x(t)}, \tag{28}$$

   where x and $\hat{x}$ are evaluated at the steady-state point. Therefore, the SSPF determines the accuracy of a state observer at the steady state [44].

2. Integral Distance Absolute error (IDAE), which is inspired by the criteria shown in Reference [44]. It measures the observer dynamic performance and penalises the errors that persist for long position. In this work, the independent variable is not the time, but it is the length; therefore, the definition is the following:

$$IDAE = \int_0^L L\,|x(L) - \hat{x}(L)|\,dL. \tag{29}$$

To evaluate the effect of the β of (16) parameter on state estimation (see to Figure 4), we propose using IDAE and SSPF (see to Figure 5). We can appreciate that, in the presence of noise, a minor fractional-order redeems the value of the IDAE. In the case of classic mode absence, the IDAE is duplicated, proving the superiority of the observer proposed in this work over the classic one.

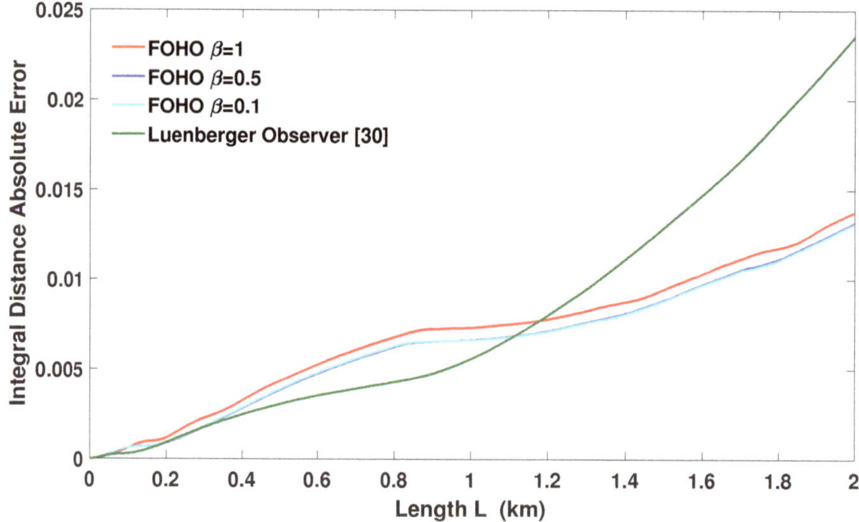

Figure 4. Observers Integral time absolute error (IDAE).

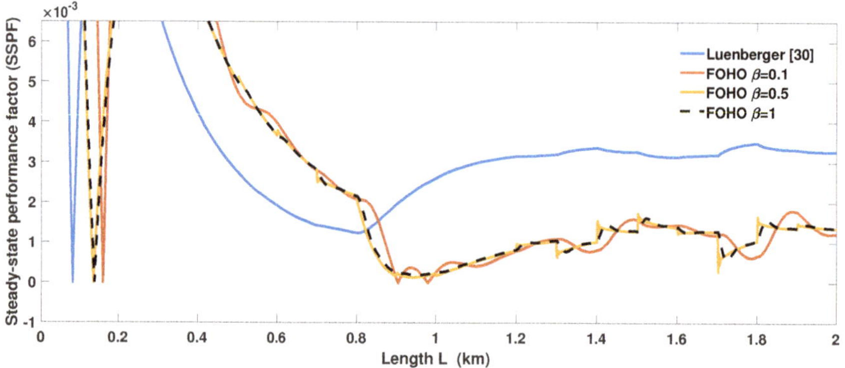

Figure 5. Observers Steady-state performance factor.

In the case of the SSPF, it appreciated a slight effect that has the fractional-order value (β), and the value of 0.1 is the one that maintains the best performance, while the classic Luenberger extended has the lowest performance. IDAE criterion importance using specific case is that the independent variable is the L length, and this criterion reflects the net yield from the river length to measure. Before reaching the steady-state, even the extended Luenberger observer [32] maintains a better performance. After $L > 1$, when the steady-state is reached, it loses effectiveness due to the parameter changes k_1 and k_2 affecting the final value of the steady-state SSPF results, as in Figure 5.

The latest generation of fractional gain observers are superior to the Luenberger Extended versions, in the presence of perturbations at the kinetic constants k_1 and k_2. However, high-gain observers depend mainly on the structure of the mathematical model. If there is a change in structure, it will result in a type of undefined perturbation, which could not be compensated by the estimator proposed in this paper. Thus, if maximum distortion is greater than the state's maximum, our algorithm would have problems because it is not a fulfilled primary theorem condition, which would be a problem or weakness of our algorithm. Therefore, it could be said that any disturbance that makes the system not comply with Conditions 2 and 3 will make the Fractional observer provided in this paper not perform correctly. When perturbations in parameters, such as U (volumetric flow rate), occur, as long as the condition shown is met, the fractional observer will maintain an excellent performance, since the condition of asymptotic stability of the main theorem is already fulfilled. Fortunately, the parametric and model distortions in most of the real practical cases meet this primary condition.

6. Conclusions

In this work, a stability study for Streeter–Phelps model water quality parameters was proposed, where it could be obtained by using Lyapunov functions. The main proposal of this work was to design a new fractional observer to estimate BOD and compare it with the classic form of an extended Luenberger observer. It has been shown that the use of FOHO improves state estimation performance and that the fractional-order plays a critical role in evaluation by analysing the error by the IDAE criteria. This type of algorithm can be used and programmed in waterborne autonomous vehicles to estimate real-time BOD and monitor DO in case of distortions along the water source position. Our laboratory is currently working on this task.

Future Work Perspectives

The main contribution presented in this work is to provide an algorithm that estimates variables, such as BOD, since there are no on-line sensors for this type of variable [26,47]. This estimating algorithm (dynamic observers) in a robotic water vehicle could be programmed, and, then, the boat robot must include a position sensing unit; some works have shown the ability to have autonomy in the position control via GPS signal [26,48]

The vehicle maintains an automatic GPS position control (L longitude) that allows it to keep a constant speed (U volume/time), similar to those of the river (U_{river} volume/time), as well as an intelligent system of obstacle avoidance that allows the vehicle to stay in the average radius of the river width; hence, $U = U_{river}$, see Figure 6. In addition, we propose that these results be presented in a comparative study with renowned modern, robust techniques [29,30] where the performance at parametric and river spatial velocity presence is studied.

Figure 6. Outline of the application of environmental monitoring.

Author Contributions: Conceptualization, A.E.R.-M.; methodology, V.G.-H., P.A.L.-P. and O.H.-G.; writing—original draft preparation, Y.B.-T. and L.E.A.-S. All authors have read and agreed to the published version of the manuscript.

Funding: This research received no external funding.

Conflicts of Interest: The authors declare no conflict of interest.

References

1. Finger, M.; Princen, T. *Environmental NGOs in World Politics: Linking the Local and the Global*; Routledge: Abingdon, UK, 2013.
2. Wang, J.; Da, L.; Song, K.; Li, B.L. Temporal variations of surface water quality in urban, suburban and rural areas during rapid urbanization in Shanghai, China. *Environ. Pollut.* **2008**, *152*, 387–393. [CrossRef] [PubMed]
3. Mendivil-Garcia, K.; Amabilis-Sosa, L.E.; Rodríguez-Mata, A.E.; Rangel-Peraza, J.G.; Gonzalez-Huitron, V.; Cedillo-Herrera, C.I.G. Assessment of intensive agriculture on water quality in the Culiacan River basin, Sinaloa, Mexico. *Environ. Sci. Pollut. Res.* **2020**. [CrossRef] [PubMed]
4. Vellidis, G.; Barnes, P.; Bosch, D.; Cathey, A. Mathematical simulation tools for developing dissolved oxygen TMDLs. *Trans. ASABE* **2006**, *49*, 1003–1022. [CrossRef]
5. Chinh, L.V.; Hiramatsu, K.; Harada, M.; Cuu, N.; Lan, T.T. Estimation of Water Environment Capacity in the Cau River Basin, Vietnam using the Streeter–Phelps Model. *J. Fac. Agric. Kyushu Univ.* **2017**, *62*, 163–169.
6. Long, B.T. Inverse algorithm for Streeter–Phelps equation in water pollution control problem. *Math. Comput. Simul.* **2020**, *171*, 119–126. [CrossRef]
7. Chii, P.L.; Rahman, H.A. Application of water quality models to rivers in Johor. In *AIP Conference Proceedings*; AIP Publishing LLC: Melville, NY, USA, 2017; Volume 1870, p. 060014.
8. Mendes, T.A.; Alves, T.L.; Monteiro, P.R.A. Modelo Streeter-Phelps para estimativa do oxigênio dissolvido em trecho urbano do rio Meia Ponte. *Tecnia* **2019**, *4*, 60–77.
9. Yustiani, Y.M.; Wahyuni, S.; Dewi, S.N.F. Determination of maximum bod load using water quality modeling of upstream citarum river. *Int. J. GEOMATE* **2019**, *16*, 118–122. [CrossRef]
10. Alva Saldaña, G.S.; Rojas Gonzales, J.A. Estimación del Déficit de Oxígeno Disuelto Usando el Modelo Streeter y Phelps en la Cuenca Baja del Río Moche. Available online: http://repositorio.ucv.edu.pe/handle/20.500.12692/40048 (accessed on 13 July 2020).
11. Rodríguez-Mata, A.E.; Amabilis-Sosa, L.E.; Roé-Sosa, A.; Barrera-Andrade, J.M.; Rangel-Peraza, J.G.; Salinas-Juárez, M.G. Quantification of recalcitrant organic compounds during their removal test by a novel and economical method based on chemical oxygen demand analysis. *Korean J. Chem. Eng.* **2019**, *36*, 423–432. [CrossRef]
12. Melo, M.; Mota, F.; Albuquerque, V.; Alexandria, A. Development of a robotic airboat for online water quality monitoring in lakes. *Robotics* **2019**, *8*, 19. [CrossRef]
13. Jorge, V.A.; Granada, R.; Maidana, R.G.; Jurak, D.A.; Heck, G.; Negreiros, A.P.; Dos Santos, D.H.; Gonçalves, L.M.; Amory, A.M. A survey on unmanned surface vehicles for disaster robotics: Main challenges and directions. *Sensors* **2019**, *19*, 702. [CrossRef]
14. Jo, W.; Hoashi, Y.; Aguilar, L.L.P.; Postigo-Malaga, M.; Garcia-Bravo, J.M.; Min, B.C. A low-cost and small USV platform for water quality monitoring. *HardwareX* **2019**, *6*, e00076. [CrossRef]
15. Nikitakos, N.; Tsaganos, G.; Papachristos, D. Autonomous Robotic Platform in Harm Environment Onboard of Ships. *IFAC-PapersOnLine* **2018**, *51*, 390–395. [CrossRef]
16. Aldegheri, S.; Bloisi, D.D.; Blum, J.J.; Bombieri, N.; Farinelli, A. Fast and power-efficient embedded software implementation of digital image stabilization for low-cost autonomous boats. In *Field and Service Robotics*; Springer: Berlin/Heidelberg, Germany, 2018; pp. 129–144.
17. Kumar, S.; Kumar, H.; Kumar, K.; Hameed, S.; Fatima, K. Real Time Water Quality Monitoring Boat. *Multidiscip. Digit. Publ. Inst. Proc.* **2018**, *2*, 1279.
18. Zappalà, G. A software set for environment monitoring networks. *WIT Trans. Ecol. Environ.* **2004**, *69*, 1–10.

19. Zappala, G.; Alberotanza, L.; Crisafi, E. Assessment of environmental conditions using automatic monitoring systems. In Proceedings of the Oceans' 99. MTS/IEEE. Riding the Crest into the 21st Century. Conference and Exhibition. Conference Proceedings (IEEE Cat. No. 99CH37008), Seattle, WA, USA, 13–16 September 1999; Volume 2, pp. 796–800.
20. Zappalà, G. A versatile software-hardware system for environmental data acquisition and transmission. *WIT Trans. Model. Simul.* **2009**, *48*, 283–294.
21. Luenberger, D. An introduction to observers. *IEEE Trans. Autom. Control* **1971**, *16*, 596–602. [CrossRef]
22. Zeitz, M. The extended Luenberger observer for nonlinear systems. *Syst. Control Lett.* **1987**, *9*, 149–156. [CrossRef]
23. Zhu, F. The design of full-order and reduced-order adaptive observers for nonlinear systems. In Proceedings of the 2007 IEEE International Conference on Control and Automation, Guangzhou, China, 30 May–1 June 2007; pp. 529–534.
24. Davila, J.; Fridman, L.; Levant, A. Second-order sliding-mode observer for mechanical systems. *IEEE Trans. Autom. Control.* **2005**, *50*, 1785–1789. [CrossRef]
25. Robles-Magdaleno, J.; Rodríguez-Mata, A.; Farza, M.; M'Saad, M. A filtered high gain observer for a class of non uniformly observable systems–Application to a phytoplanktonic growth model. *J. Process Control* **2020**, *87*, 68–78. [CrossRef]
26. Cao, Z.; Zhang, R.; Yang, Y.; Lu, J.; Gao, F. Discrete-time robust iterative learning Kalman filtering for repetitive processes. *IEEE Trans. Autom. Control.* **2015**, *61*, 270–275. [CrossRef]
27. Torres, L.; Jiménez-Cabas, J.; Gómez-Aguilar, J.F.; Pérez-Alcazar, P. A simple spectral observer. *Math. Comput. Appl.* **2018**, *23*, 23. [CrossRef]
28. Rodríguez-Mata, A.E.; González-Hernández, I.; Rangel-Peraza, J.G.; Salazar, S.; Leal, R.L. Wind-gust compensation algorithm based on high-gain residual observer to control a quadrotor aircraft: Real-time verification task at fixed point. *Int. J. Control. Autom. Syst.* **2018**, *16*, 856–866. [CrossRef]
29. Gómez-Peñate, S.; Valencia-Palomo, G.; López-Estrada, F.R.; Astorga-Zaragoza, C.M.; Osornio-Rios, R.A.; Santos-Ruiz, I. Sensor fault diagnosis based on a sliding mode and unknown input observer for Takagi-Sugeno systems with uncertain premise variables. *Asian J. Control.* **2019**, *21*, 339–353. [CrossRef]
30. López-Estrada, F.R.; Theilliol, D.; Astorga-Zaragoza, C.M.; Ponsart, J.C.; Valencia-Palomo, G.; Camas-Anzueto, J. Fault diagnosis observer for descriptor Takagi-Sugeno systems. *Neurocomputing* **2019**, *331*, 10–17. [CrossRef]
31. Alegria-Zamudio, M.; Escobar-Jiménez, R.; Gómez-Aguilar, J.; García-Morales, J.; Hernández-Pérez, J. Double pipe heat exchanger temperatures estimation using fractional observers. *Eur. Phys. J. Plus* **2019**, *134*, 496. [CrossRef]
32. Vorce, T.; Mulholland, R. Estimating Oxygen demand in aquatic ecosystems. In Proceedings of the ICASSP'78. IEEE International Conference on Acoustics, Speech, and Signal Processing, Tulsa, OK, USA, 10–12 April 1978; Volume 3, pp. 402–404.
33. Rinaldi, S.; Soncini-Sessa, R. Sensitivity analysis of generalized Streeter-Phelps models. *Adv. Water Resour.* **1978**, *1*, 141–146. [CrossRef]
34. Gotovtsev, A. Modification of the Streeter-Phelps system with the aim to account for the feedback between dissolved oxygen concentration and organic matter oxidation rate. *Water Resour.* **2010**, *37*, 245–251. [CrossRef]
35. Chapra, S.C. *Surface Water-Quality Modeling*; Waveland Press: Long Grove, IL, USA, 2008.
36. Khalil, H.K.; Grizzle, J.W. *Nonlinear Systems*; Prentice hall: Upper Saddle River, NJ, USA, 2002; Volume 3.
37. Rodríguez-Mata, A.E.; Luna, R.; Pérez-Correa, J.R.; Gonzalez-Huitrón, A.; Castro-Linares, R.; Duarte-Mermoud, M.A. Fractional Sliding Mode Nonlinear Procedure for Robust Control of an Eutrophying Microalgae Photobioreactor. *Algorithms* **2020**, *13*, 50. [CrossRef]
38. Tepljakov, A.; Petlenkov, E.; Belikov, J. Gain and order scheduled fractional-order PID control of fluid level in a multi-tank system. In Proceedings of the ICFDA'14 International Conference on Fractional Differentiation and Its Applications 2014, Catania, Italy, 23–25 June 2014; pp. 1–6.
39. Celikovsky, S.; Torres-Muñoz, J.; Rodriguez-Mata, A.; Dominguez-Bocanegra, A.R. An adaptive extension to high gain observer with application to wastewater monitoring. In Proceedings of the 2015 12th International Conference on Electrical Engineering, Computing Science and Automatic Control (CCE), Mexico City, Mexico, 28–30 October 2015; pp. 1–6.

40. Govea-Vargas, A.; Castro-Linares, R.; Duarte-Mermoud, M.A.; Aguila-Camacho, N.; Ceballos-Benavides, G.E. Fractional order sliding mode control of a class of second order perturbed nonlinear systems: Application to the trajectory tracking of a quadrotor. *Algorithms* **2018**, *11*, 168. [CrossRef]
41. Izaguirre-Fierro, G.; Páez-Osuna, F.; Osuna-López, J. Heavy metals in fishes from Culiacan valley, Sinaloa, Mexico. *Cienc. Mar.* **1992**, *18*, 143–151. [CrossRef]
42. Ruiz-Fernández, A.C.; Hillaire-Marcel, C.; Ghaleb, B.; Soto-Jiménez, M.; Páez-Osuna, F. Recent sedimentary history of anthropogenic impacts on the Culiacan River Estuary, northwestern Mexico: Geochemical evidence from organic matter and nutrients. *Environ. Pollut.* **2002**, *118*, 365–377. [CrossRef]
43. Ruiz-Fernández, A.; Hillaire-Marcel, C.; Páez-Osuna, F.; Ghaleb, B.; Soto-Jiménez, M. Historical trends of metal pollution recorded in the sediments of the Culiacan River Estuary, Northwestern Mexico. *Appl. Geochem.* **2003**, *18*, 577–588. [CrossRef]
44. Olanrewaju, M.J.; Al-Arfaj, M.A. Design and implementation of state observers in reactive distillation. *Chem. Eng. Commun.* **2007**, *195*, 267–292. [CrossRef]
45. Oliveira, R.; Ferreira, E.; Oliveira, F.; De Azevedo, S.F. A study on the convergence of observer-based kinetics estimators in stirred tank bioreactors. *J. Process Control* **1996**, *6*, 367–371. [CrossRef]
46. Paesa, D.; Franco, C.; Llorente, S.; Lopez-Nicolas, G.; Sagues, C. Reset observers applied to MIMO systems. *J. Process Control* **2011**, *21*, 613–619. [CrossRef]
47. Wang, K.; Lu, H.; Liu, B.; Yang, J. A high-efficiency and low-cost AEE polyurethane chemo-sensor for Fe3+ and explosives detection. *Tetrahedron Lett.* **2018**, *59*, 4191–4195. [CrossRef]
48. Wendeborn, D.; Armstrong, R.; Nowicki, S.; Hill, S. Method for the Automated Docking of Robotic Platforms. U.S. Patent App. 16/669,363, 30 October 2020.

© 2020 by the authors. Licensee MDPI, Basel, Switzerland. This article is an open access article distributed under the terms and conditions of the Creative Commons Attribution (CC BY) license (http://creativecommons.org/licenses/by/4.0/).

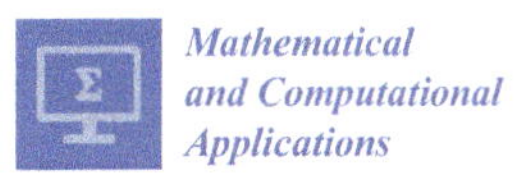

Article

Robust qLPV Tracking Fault-Tolerant Control of a 3 DOF Mechanical Crane

Francisco-Ronay López-Estrada [1], Oscar Santos-Estudillo [1], Guillermo Valencia-Palomo [2,*], Samuel Gómez-Peñate [1] and Carlos Hernández-Gutiérrez [1]

1 TURIX-Dynamics Diagnosis and Control Group, Tecnológico Nacional de México, IT Tuxtla Gutiérrez, Carretera Panamericana Km 1080, C.P. 29050 Tuxtla Gutiérrez, Chiapas, Mexico; frlopez@ittg.edu.mx (F.-R.L.-E.); ossantos1994@gmail.com (O.S.-E.); sgomez@ittg.edu.mx (S.G.-P.); postgraduatecahg@gmail.com (C.H.-G.)

2 Tecnológico Nacional de México, IT Hermosillo, Ave. Tecnológico y Periférico Poniente S/N, C.P. 83170 Hermosillo, Sonora, Mexico

* Correspondence: gvalencia@hermosillo.tecnm.mx

Received: 5 July 2020; Accepted: 28 July 2020; Published: 28 July 2020

Abstract: The main aim of this paper is to propose a robust fault-tolerant control for a three degree of freedom (DOF) mechanical crane by using a convex quasi-Linear Parameter Varying (qLPV) approach for modeling the crane and a passive fault-tolerant scheme. The control objective is to minimize the load oscillations while the desired path is tracked. The convex qLPV model is obtained by considering the nonlinear sector approach, which can represent exactly the nonlinear system under the bounded nonlinear terms. To improve the system safety, tolerance to partial actuator faults is considered. Performance requirements of the tracking control system are specified in an $\mathcal{H}_\infty$ criteria that guarantees robustness against measurement noise, and partial faults. As a result, a set of Linear Matrix Inequalities is derived to compute the controller gains. Numerical experiments on a realistic 3 DOF crane model confirm the applicability of the control scheme.

Keywords: 3 DOF crane; convex systems; fault-tolerant control; robust control; qLPV systems; Takagi–Sugeno systems

1. Introduction

In recent years, fault-tolerant control (FTC) has become a relevant research field and has attracted significant attention because of its applicability to industrial systems, which increases their security and reliability. A fault can be defined as abnormal behavior of at least one characteristic property or parameter that changes the system performance [1,2]. It is important to note that a fault denotes a breakdown rather than a catastrophe [3,4]. In other words, a fault not necessarily ends in a system stop. However, if no action is taken on time, the system performance begins to degrade that could end in a catastrophe [5]. Therefore, in order to guarantee a minimum level of performance, it is necessary to develop methods to improve system safety and reliability.

Model-based safety schemes require differential equations representing the complex dynamics presented in physical systems, which are often nonlinear [6,7]. Recently, multimodel techniques such as Linear Parameter Varying (LPV), quasi-LPV (qLPV), and Takagi–Sugeno (TS) systems have emerged as an attractive alternative to deal with the analysis of complex nonlinear systems due to the fact that it is possible to extend techniques developed for linear systems but applied to nonlinear systems [8–11]. In this paper, it is considered that qLPV and TS systems are the same because the convex model is obtained through the so-called nonlinear sector approach [12]. This paradigm has been extensively studied in the works of [13,14]. In the literature, FTC systems have been classified into two approaches:

passive and active [15]. Passive FTC is an extension of robust control [16] and requires some knowledge of possible failures that may affect the system. In this scheme, the controller is designed a priori to be robust to faults, and non-online adaptation is made. This type of control is interesting because it does not need any fault diagnosis module [17]. In contrast, active FTC systems offer flexibility in the design task. They are assimilated as a variable structure technique because the controller is reconfigured when a fault occurs [18]. However, it is necessary to include a Fault Diagnosis and Isolation (FDI) module, which provides information about the faults [19,20]. The inclusion of the FDI module gives some conservatism into the controller solution. This work is dedicated to the study of passive fault-tolerant with application to a mechanical crane.

A mechanical crane can load hundreds of tons and are widely used in oil platforms, ships, factories, railway depots, piers, among others [21]. By design, the crane is a sub-actuated system, which means that it has more degrees of freedom (DOF) than control inputs. In the particular case of the crane shown in Figure 1, it is assumed that the load is attached to a plane. The degrees of freedom of the crane are three: the first one refers to movement on the x-axis (forward/backward movement of the carriage); the second on the z-axis (up/down movement of the load); and finally the angular displacement of the load on the x-axis. However, the system has only two actuators, which are the trolley motor and the hoist motor. The control objective is to locate the crane at the desired position, whereby the trolley motor moves the load as fast as possible. At the same time, the oscillations that could destabilize the system must be minimized [22]. Due to this, it is essential to design control schemes that consider the under-actuation, a large number of linearities, possible faults, and robustness to the payload oscillations.

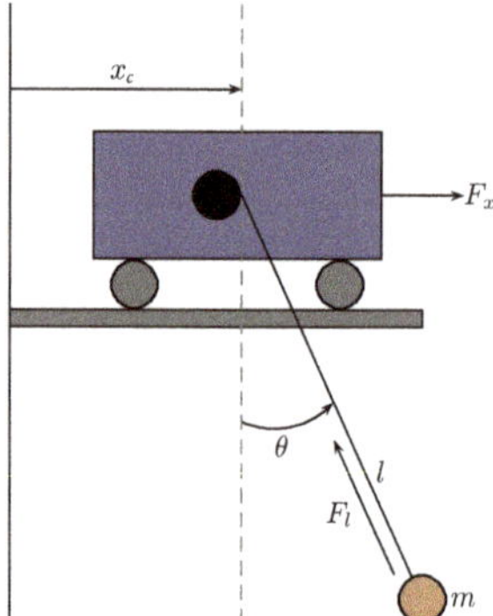

Figure 1. Three degrees of freedom mechanical crane.

In the literature, this problem has been approached by considering different methods, e.g., sliding modes [23], LQ controllers [24], particle swarm optimization [25], adaptive control [26,27], among others. Regarding safety systems, sliding mode differentiators can be consulted in [28] and fault-tolerant control in [29,30]. Some authors have explored the design of robust controllers based on multimodel techniques. For example, in [31], a robust stabilization method based on TS models was proposed. The authors in [32] presented a hybrid controller that includes position regulation and oscillation control designed using TS techniques. Similarly, in [33], a reduced-order $\mathcal{H}_2/\mathcal{H}_\infty$ LPV controller was proposed. In [34] a distributed parallel compensation control is explored through TS techniques. In [35], a fault-tolerant LPV control was proposed. Authors in [36] proposed a reconfiguration scheme for active fault tolerance by considering predictive control. However, none of the reported works were subjected to performance tests in the presence of disturbances by unknown additive signals.

This paper is devoted to developing a robust quasi-Linear Parameter Varying (qLPV) fault tolerant-control system with $\mathcal{H}_\infty$ criteria applied to a 3 DOF crane. The main idea is to propose a control law that minimize oscillations in the load while the desired path is tracked. The proposed method is robust to disturbances, sensor noise, and partial faults. The method performance and applicability are tested through numerical simulations on a 3 DOF mechanical crane by compensating partial faults on the trolley and load motors.

2. Mathematical Modeling

The free-body diagram of three degrees of freedom traveling crane is shown in Figure 1. The cart slides along a horizontal rail and at the same time that it is supported by two metal legs. A mass is suspended from a cable attached to the cart. Force F_x is applied to the cart, which provokes a displacement in the x-axis. This movement causes an angular displacement θ formed in the pendulum by the load mass m and the cable of length l whose value can be changed by F_l, i.e., F_l activates the elevation system by means of a motor and gears. The nonlinear model is given by the following ordinary differential equations [23]:

$$M(q)\ddot{q} + D\dot{q} + C(q,\dot{q})\dot{q} + G(q) = F; \tag{1}$$

with:

$$q = \begin{bmatrix} x_c(t) \\ l(t) \\ \theta(t) \end{bmatrix}; F = \begin{bmatrix} F_x(t) \\ F_l(t) \\ 0 \end{bmatrix}; M(q) = \begin{bmatrix} (M_x + m) & m\sin\theta(t) & ml(t)\cos\theta(t) \\ m\sin\theta(t) & (M_l + m) & 0 \\ ml(t)\cos\theta(t) & 0 & ml(t)^2 \end{bmatrix};$$

$$C(q,\dot{q}) = \begin{bmatrix} 0 & 2m\cos\theta(t)\dot{\theta}(t) & -ml\sin\theta(t)\dot{\theta}(t) \\ 0 & 0 & -ml(t)\dot{\theta}(t) \\ 0 & 2m\dot{\theta}(t)l(t) & 0 \end{bmatrix}; D = \begin{bmatrix} D_x & 0 & 0 \\ 0 & D_l & 0 \\ 0 & 0 & 0 \end{bmatrix}; G(q) = \begin{bmatrix} 0 \\ mg - mg\cos\theta(t) \\ mgl(t)\sin\theta(t) \end{bmatrix};$$

where m is the load mass and g is the gravitational acceleration; D_x and D_y are the viscous damping coefficients associated with the x- and z-axis respectively; finally, M_x and M_l are the traveling and hoisting components of the crane mass, respectively, i.e., $M_x = m + m_c$ and $M_l = m$, where m_c is the cart mass. The parameter values considered in this paper are close to a real laboratory crane system, e.g., [37].

The state-space derivation is obtained solving (1) for $\ddot{q}$, such as:

$$\ddot{q} = -M(q)^{-1}D\dot{q} - M(q)^{-1}C(q,\dot{q})\dot{q} - M(q)^{-1}G(q) + M(q)^{-1}F. \tag{2}$$

By using $\sin\theta(t) = S_\theta$, and $\cos\theta(t) = C_\theta$; Equation (2) in long hand becomes,

$$\ddot{q} = -\begin{bmatrix} 0 & 0 & -\dfrac{mM_lS_\theta l(t)\dot{\theta}(t)}{\mu(t)} \\ 0 & 0 & -\dfrac{mM_xl(t)\dot{\theta}(t)}{\mu(t)} \\ 0 & \dfrac{2\dot{\theta}(t)}{l(t)} & \dfrac{M_lmC_\theta S_\theta\dot{\theta}(t)}{\mu(t)} \end{bmatrix}\dot{q} - \begin{bmatrix} \dfrac{(M_l+m)D_x}{\mu(t)} & -\dfrac{mS_\theta D_l}{\mu(t)} & 0 \\ -\dfrac{mS_\theta D_x}{\mu(t)} & \dfrac{(M_x+m-mC_\theta^2)D_l}{\mu(t)} & 0 \\ -\dfrac{C_\theta(M_l+m)D_x}{l(t)\mu(t)} & \dfrac{mC_\theta S_\theta D_l}{l(t)\mu(t)} & 0 \end{bmatrix}\dot{q}$$

$$-\begin{bmatrix} \dfrac{-mgS_\theta(t)(M_lC_\theta+m)}{\mu(t)} & 0 \\ 0 & \dfrac{(M_x+m+mC_\theta^2)(mg-mgC_\theta)+gm^2C_\theta S_\theta^2}{\mu(t)} \\ 0 & 0 \end{bmatrix.$$

$$\begin{matrix} 0 \\ 0 \\ \dfrac{mC_\theta S_\theta(mg-mgC_\theta)+(M_xM_l+M_lm+M_xm+m^2-m^2S_\theta^2)gS_\theta}{l(t)\mu(t)} \end{matrix}\Bigg]$$

$$+\begin{bmatrix} \frac{M_l+m}{\mu(t)} & -\frac{mS_\theta}{\mu(t)} & -\frac{(M_l+m)C_\theta}{l(t)\mu(t)} \\ -\frac{mS_\theta}{\mu(t)} & \frac{M_x+m-mC_\theta^2}{\mu(t)} & \frac{mS_\theta C_\theta}{l(t)\mu(t)} \\ -\frac{(M_l+m)C_\theta}{l(t)\mu(t)} & \frac{mS_\theta C_\theta}{l(t)\mu(t)} & \frac{M_lM_x+M_lm+M_xm+m^2-m^2S_\theta^2}{m(l(t))^2\mu(t)} \end{bmatrix}\begin{bmatrix} F_x(t) \\ F_l(t) \\ 0 \end{bmatrix};$$

with $\mu(t)=M_lM_x+M_lm+M_xm+m^2-M_lmC_\theta^2-m^2C_\theta^2-m^2S_\theta^2=M_lM_x+M_lm+M_xm-M_lmC_\theta^2$. Recalling that $q=[x_c(t),l(t),\theta(t)]^T$; then, by considering small values of θ, i.e., $S_\theta\approx\theta$, $C_\theta\approx 1$, $\theta^2\approx 0$, and $\dot{\theta}^2\approx 0$, the following equations can be obtained,

$$\begin{aligned} \ddot{x}_c(t) &= -\frac{(M_l+m)D_x\dot{x}_c(t)}{M_lM_x+M_xm}+\frac{m\theta(t)D_l\dot{l}(t)}{M_lM_x+M_xm}+\frac{(M_l+m)mg\theta(t)}{M_lM_x+M_xm}+\frac{(M_l+m)F_x(t)}{M_lM_x+M_xm}-\frac{m\theta(t)F_l(t)}{M_lM_x+M_xm}; \\ \ddot{l}(t) &= \frac{mD_x\theta(t)\dot{x}_c(t)}{M_lM_x+M_xm}-\frac{M_xD_l\dot{l}(t)}{M_lM_x+M_xm}-\frac{m\theta(t)F_x(t)}{M_lM_x+M_xm}+\frac{M_xF_l(t)}{M_lM_x+M_xm}; \\ \ddot{\theta}(t) &= \frac{(M_l+m)D_x\dot{x}_c(t)}{l(t)(M_lM_x+M_xm)}-\frac{mD_l\theta(t)\dot{l}(t)}{l(t)(M_lM_x+M_xm)}-\frac{2\dot{\theta}(t)\dot{l}(t)}{l(t)}-\frac{(M_xM_l+M_lm+M_xm+m^2)g\theta(t)}{l(t)(M_lM_x+M_xm)} \\ &\quad -\frac{(M_l+m)F_x(t)}{l(t)(M_lM_x+M_xm)}+\frac{m\theta F_l(t)}{l(t)(M_lM_x+M_xm)}. \end{aligned} \tag{3}$$

Remark 1. *If $\theta(t)$ is not assumed to be small, the nonlinear terms would increase, increasing the complexity of the convex system representation resulting in a more involved procedure for the controller design. However, the consideration that the designed controller will keep load oscillations small makes it possible to assume $\theta\approx 0$.*

Finally, by setting $x(t)=[x_1(t), x_2(t), x_3(t), x_4(t), x_5(t), x_6(t)]^T=[x_c(t),\dot{x}_c(t),l(t),\dot{l}(t),\theta(t),\dot{\theta}(t)]^T$ and $u(t)=[F_x(t),F_l(t)]^T$, the state space representation is obtained:

$$\dot{x}(t)=\begin{bmatrix} 0 & 1 & 0 & 0 & 0 & 0 \\ 0 & -m_2D_x & 0 & m_1D_lx_5 & m_4g & 0 \\ 0 & 0 & 0 & 1 & 0 & 0 \\ 0 & m_1D_xx_5 & 0 & -m_3D_l & 0 & 0 \\ 0 & 0 & 0 & 0 & 0 & 1 \\ 0 & \frac{m_2D_x}{x_3} & 0 & -(\frac{m_1D_lx_5}{x_3}+\frac{2x_6}{x_3}) & \frac{-m_5g}{x_3} & 0 \end{bmatrix}x(t)+\begin{bmatrix} 0 & 0 \\ m_2 & -m_1x_5 \\ 0 & 0 \\ -m_1x_5 & m_3 \\ 0 & 0 \\ -\frac{m_2}{x_3} & \frac{m_1x_5}{x_3} \end{bmatrix}u(t); \tag{4}$$

with $m_1=m/(M_lM_x+M_xm)$; $m_2=(M_l+m)/(M_lM_x+M_xm)$; $m_3=M_x/(M_lM_x+M_xm)$; $m_4=m(M_l+m)/(M_lM_x+M_xm)$; $m_5=(M_xM_l+M_lm+M_xm+m^2)/(M_lM_x+M_xm)$. Note that the nonlinear terms in (4) are given by:

$$z=\begin{bmatrix} x_5 & \frac{1}{x_3} & \frac{x_5}{x_3} & \frac{x_6}{x_3} \end{bmatrix}.$$

In order to obtain a TS model through the nonlinear sector approach, each nonlinear term is bounded as $x_3\in[0.1,0.72]$ [m], $x_5\in[-0.35,0.35]$ [rad], and $x_6\in[-3.467,3.467]$ [rad/s], such as the weighting functions are described as:

1. For $z_1=x_5$ the limits are $z_{1,min}=-0.35$ and $z_{1,max}=0.35$. The weigthing functions are $w_{11}=\frac{z_{1,max}-z_1}{z_{1,max}-z_{1,min}}$ and $w_{12}=1-w_{11}$. Therefore, z_1 can be rewritten as $z_1=z_{1,min}w_{11}+z_{1,max}w_{12}$.
2. For $z_2=\frac{1}{x_3}$ the limits are $z_{2,min}=1$ and $z_{2,max}=10$. The weigthing functions are $w_{21}=\frac{z_{2,max}-z_2}{z_{2,max}-z_{1,min}}$ and $w_{22}=1-w_{21}$. Therefore, $z_2(t)$ can be rewritten as $z_2=z_{2,min}w_{21}+z_{2,max}w_{22}$.
3. $z_3=\frac{x_5}{x_3}$ is bounded as $z_{3,min}=-3.5$ and $z_{3,max}=3.5$. The weigthing functions are $w_{31}=\frac{z_{3,max}-z_3}{z_{3,max}-z_{3,min}}$ and $w_{32}=1-w_{31}$, with $z_3=z_{3,min}w_{31}+z_{3,max}w_{32}$.

4. Finally, $z_4 = \frac{x_6}{x_3}$ is bounded as $z_{4,min} = -34.67$ and $z_{4,max} = 34.67$. The weighting functions are $w_{41} = \frac{z_{4,max} - z_4}{z_{4,max} - z_{4,min}}$ and $w_{42} = 1 - w_{41}$, with $z_4 = z_{4,min} w_{41} + z_{4,max} w_{42}$.

Then, the nonlinear model can be represented by:

$$\dot{x}(t) = \begin{bmatrix} 0 & 1 & 0 & 0 & 0 & 0 \\ 0 & -m_2 D_x & 0 & m_1 D_l z_1 & m_4 g & 0 \\ 0 & 0 & 0 & 1 & 0 & 0 \\ 0 & m_1 D_x z_1 & 0 & -m_3 D_l & 0 & 0 \\ 0 & 0 & 0 & 0 & 0 & 1 \\ 0 & m_2 D_x z_2 & 0 & -(m_1 D_l z_3 + 2z_4) & -m_5 g z_2 & 0 \end{bmatrix} x(t) + \begin{bmatrix} 0 & 0 \\ m_2 & -m_1 z_1 \\ 0 & 0 \\ -m_1 z_1 & m_3 \\ 0 & 0 \\ -m_2 z_2 & m_1 z_3 \end{bmatrix} u(t) \tag{5}$$

The number of local sub-models is $2^4 = 16$, then, the membership functions are computed as the product of the weighting functions that correspond to each local model,

$$h_i(z(t)) = \prod_{j=1}^{p} w_{ij}^{j}(z_j), \quad i = 1, 2, \ldots, 2^p. \tag{6}$$

Note that the membership functions are convex which means that $h_i(z(t)) \geq 0$, $\sum_{i=1}^{16} h_i(z(t)) = 1$. The number of combinations are defined as given in Table 1.

Table 1. Weighing functions.

h_i	Combination	h_i	Combination
h_1	$w_0^1 w_0^2 w_0^3 w_0^4$	h_9	$w_1^1 w_0^2 w_0^3 w_0^4$
h_2	$w_0^1 w_0^2 w_0^3 w_1^4$	h_{10}	$w_1^1 w_0^2 w_0^3 w_1^4$
h_3	$w_0^1 w_0^2 w_1^3 w_0^4$	h_{11}	$w_1^1 w_0^2 w_1^3 w_0^4$
h_4	$w_0^1 w_0^2 w_1^3 w_1^4$	h_{12}	$w_1^1 w_0^2 w_1^3 w_1^4$
h_5	$w_0^1 w_1^2 w_0^3 w_0^4$	h_{13}	$w_1^1 w_1^2 w_0^3 w_0^4$
h_6	$w_0^1 w_1^2 w_0^3 w_1^4$	h_{14}	$w_1^1 w_1^2 w_0^3 w_1^4$
h_6	$w_0^1 w_1^2 w_1^3 w_0^4$	h_{14}	$w_1^1 w_1^2 w_1^3 w_0^4$
h_7	$w_0^1 w_1^2 w_1^3 w_0^4$	h_{15}	$w_1^1 w_1^2 w_1^3 w_0^4$
h_8	$w_0^1 w_1^2 w_1^3 w_1^4$	h_{16}	$w_1^1 w_1^2 w_1^3 w_1^4$

Then, the convex qLPV model is derived as:

$$\begin{aligned} \dot{x}(t) &= \sum_{i=1}^{16} h_i(z) \left[A_i x(t) + B_i u(t) \right]; \\ y(t) &= Cx(t); \end{aligned} \tag{7}$$

The matrix C is constant as $y(t)$ represents the measured output which according to the nature of the system is linear.

3. Convex $\mathcal{H}_\infty$ Fault-Tolerant Controller

Under the presence of additive actuator faults, system (7) can be rewritten as

$$\begin{aligned} \dot{\boldsymbol{x}}(t) &= A_h \boldsymbol{x}(t) + B_h \boldsymbol{u}(t) + G_h f(t), \\ y(t) &= C\boldsymbol{x}(t), \end{aligned} \tag{8}$$

where:

$$A_h = \sum_{i=1}^{16} h_i(z(x(t)))A_i,\ B_h = \sum_{i=1}^{16} h_i(z(x(t)))B_i, \tag{9}$$

$f \in \mathbb{R}^s$ represents the additive fault vector and $G_h \in \mathbb{R}^{n\times s}$ represents the fault matrix. Typically in order to simulate actuator degradation, it is considered that $G_h = B_h$.

Then, a convex qLPV controller for the nonlinear 3 DOF crane, as the one shown in Figure 2, is proposed with:

$$\dot{\epsilon}(t) = \omega(t) - y(t) = \omega(t) - Cx(t), \tag{10}$$

where $\omega(t)$ is the reference and the control law is defined by:

$$u(t) = \sum_{i=1}^{16} h_i(z)\left[F_{1i}x(t) + F_{2i}\epsilon(t)\right] = \sum_{i=1}^{16} h_i(z)\mathcal{F}_i \begin{bmatrix} x(t) \\ \epsilon(t) \end{bmatrix} = \mathcal{F}_h \begin{bmatrix} x(t) \\ \epsilon(t) \end{bmatrix}, \tag{11}$$

where F_{1i} and F_{2i} are the gain matrices to be computed. Then, the main problem is to determine the optimal values for these control gains, such that the system be robust to disturbances and sensor noise. Then, by considering the tracking comparator in the control scheme (10), the following augmented system is obtained:

$$\dot{\bar{x}}(t) = \bar{A}_h\bar{x}(t) + \bar{B}_h u(t) + \bar{G}_h f(t) + \bar{B}_\omega \omega(t) \tag{12}$$

with:

$$\bar{A}_h = \begin{bmatrix} A_h & 0 \\ -C & 0 \end{bmatrix},\ \bar{B}_h = \begin{bmatrix} B_h \\ 0 \end{bmatrix},\ \bar{B}_\omega = \begin{bmatrix} 0 \\ I \end{bmatrix},\ \bar{G}_h = \begin{bmatrix} G_h \\ 0 \end{bmatrix},\ \bar{x} = \begin{bmatrix} x(t) \\ \epsilon(t) \end{bmatrix}.$$

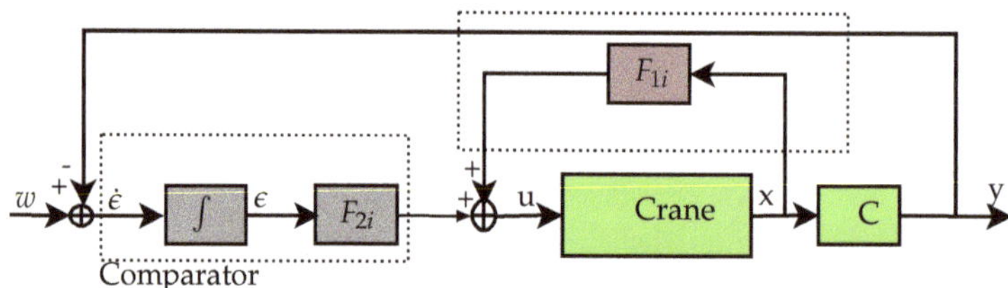

Figure 2. Convex tracking control diagram.

Assuming that the pair $[\bar{A}_i, \bar{B}_i]$ is controllable and the control law (11), the following closed-loop system is obtained:

$$\dot{\bar{x}} = (\bar{A}_h + \bar{B}_h\mathcal{F}_h)\,\bar{x} + \bar{G}_h f(t) + \bar{B}_\omega \omega(t). \tag{13}$$

System (13) can be rewritten equivalently as:

$$\dot{\bar{x}} = (\bar{A}_h + \bar{B}_h\mathcal{F}_h)\,\bar{x} + \bar{G}_{\omega h}\bar{f}_\omega(t), \tag{14}$$

with:

$$\bar{G}_{\omega h} = \begin{bmatrix} \bar{G}_h & \bar{B}_\omega \end{bmatrix},\ \bar{f}_\omega = \begin{bmatrix} f(t) \\ \omega(t) \end{bmatrix} \tag{15}$$

Then, by considering an $\mathcal{H}_\infty$ performance criteria is considered to design a robust controller $\mathcal{F}_h$, which minimizes the energy $\mathcal{L}_2$-gain of the closed-loop system, such as norm upper bound is simultaneously guaranteed:

$$\frac{\|\bar{x}(t)\|_2}{\|\bar{f}(t)_\omega\|_2} \leq \gamma, \gamma > 0,$$

as a result, the following Theorem is derived:

Theorem 1. *Given the qLPV system (7), the robust control (11) has a quadratic γ-performance level if there exist matrices X, M_j, with $\gamma > 0$, such that the following optimization problem is solved $\forall i, j \in [1, 2, \ldots, 16]$:*

$$min\ \bar{\gamma},\ s.t.$$

$$\begin{bmatrix} \bar{A}_i X + \bar{B}_i M_j + M_j^T \bar{B}_i^T + X \bar{A}_i^T & \bar{G}_{\omega i} & X^T \\ \bar{G}_{\omega i}^T & -\bar{\gamma} I & 0 \\ X & 0 & -I \end{bmatrix} \leq 0. \tag{16}$$

Then, the controller matrices and the performance are computed by $\mathcal{F}_i = M_j X^{-1}$ and $\gamma = \sqrt{\bar{\gamma}}$.

Proof. Let us consider the following $\mathcal{H}_\infty$ performance criteria:

$$\dot{V}(\bar{x}(t)) + \bar{x}(t)^T \bar{x}(t) \leq \gamma^2 \bar{f}_\omega(t)^T \bar{f}_\omega(t), \tag{17}$$

where $\dot{V}(\bar{x}(t))$ is the derivative of the quadratic Lyapunov function $\bar{x}(t)^T P \bar{x}(t) > 0$, with $P = P^T > 0$, over the trajectory of the augmented states, such as the performance criteria can be rewritten as:

$$\dot{\bar{x}}(t)^T P \bar{x}(t) + \bar{x}(t)^T P \dot{\bar{x}}(t) + \bar{x}(t)^T \bar{x}(t) - \gamma^2 \bar{f}_\omega(t)^T \bar{f}_\omega(t) \leq 0. \tag{18}$$

Then, by considering the augmented matrices given in (14), the following is obtained:

$$\bar{x}^T P((\bar{A}_h + \bar{B}_h \mathcal{F}_h)\bar{x} + \bar{G}_{\omega h} \bar{f}_\omega) + ((\bar{A}_h + \bar{B}_h \mathcal{F}_h)\bar{x} + \bar{G}_{\omega h} \bar{f}_\omega)^T P \bar{x} + \bar{x}^T \bar{x} - \gamma^2 \bar{f}_\omega^T \bar{f}_\omega \leq 0, \tag{19}$$

which can be equivalently rewritten as:

$$\bar{x}^T (P\bar{A}_h + P\bar{B}_h \mathcal{F}_h + \mathcal{F}_h^T \bar{B}_h^T P + \bar{A}_h^T P + I)\bar{x} + \bar{x}^T (P\bar{G}_{\omega h})\bar{f}_\omega + \bar{f}_\omega^T (\bar{G}_{\omega h}^T P)\bar{x} - \bar{f}_\omega^T (\gamma^2 I)\bar{f}_\omega \leq 0. \tag{20}$$

Then, the performance criteria can be factorized as:

$$\begin{bmatrix} \bar{x}^T & \bar{f}_\omega^T \end{bmatrix} \begin{bmatrix} P\bar{A}_h + P\bar{B}_h \mathcal{F}_h + \mathcal{F}_h^T \bar{B}_h^T P + \bar{A}_h^T P + I & P\bar{G}_{\omega h} \\ \bar{G}_{\omega h}^T P & -\gamma^2 I \end{bmatrix} \begin{bmatrix} \bar{x} \\ \bar{f}_\omega \end{bmatrix} \leq 0. \tag{21}$$

In order to put together the unknown matrices P and $\mathcal{F}_h$, the inequality is pre and post-multiplyied by $\begin{bmatrix} X & 0 \\ 0 & I \end{bmatrix}$ and its transpose, with $X = P^{-1}$, such as the following is obtained:

$$\begin{bmatrix} \bar{A}_h X + \bar{B}_h \mathcal{F}_h X + X \mathcal{F}_h^T \bar{B}_h^T + X \bar{A}_h^T + X^T X & \bar{G}_{\omega h} \\ \bar{G}_{\omega h}^T & -\gamma^2 I \end{bmatrix} \leq 0 \tag{22}$$

With this transformation, the quadratic term can be eliminated by considering $M_h = \mathcal{F}_h X$ and $\bar{\gamma} = \gamma^2$, such as the following Linear Matrix Inequality (LMI) is derived:

$$\begin{bmatrix} \bar{A}_h X + \bar{B}_h M_h + M_h^T \bar{B}_h^T + X \bar{A}_h^T + X^T X & \bar{G}_{\omega h} \\ \bar{G}_{\omega h}^T & -\bar{\gamma} I \end{bmatrix} \leq 0, \tag{23}$$

which can be rewritten as:

$$\begin{bmatrix} \bar{A}_h X + \bar{B}_h M_h + M_h^T \bar{B}_h^T + X \bar{A}_h^T & \bar{G}_{\omega h} \\ \bar{G}_{\omega h}^T & -\bar{\gamma} I \end{bmatrix} + \begin{bmatrix} X^T \\ 0 \end{bmatrix} I \begin{bmatrix} X & 0 \end{bmatrix} \leq 0. \tag{24}$$

Then, by considering the Schur complement, the following LMI is obtained:

$$\begin{bmatrix} \bar{A}_h X + \bar{B}_h M_h + M_h^T \bar{B}_h^T + X \bar{A}_h^T & \bar{G}_{\omega h} & X^T \\ \bar{G}_{\omega h}^T & -\bar{\gamma} I & 0 \\ X & 0 & -I \end{bmatrix} \leq 0. \tag{25}$$

Note that (24) and (25) are equivalent, and both can be used to find the gains. However, (25) is written in a relaxed form to reduce the LMI conservatism. Finally, by considering the equivalent matrices in (9), the LMI, as given in Theorem 1, is obtained. This completes the proof. □

Remark 2. *In this paper, the method considers a constant Lyapunov matrix P, which means it is necessary to find a Matrix P such for all 16 LMIs. This problem can be relaxed by considering a parameter-varying P_h, which is also called the non-quadratic Lyapunov functions [38]. This problem can reduce the conservatism of the LMI and open new research areas for future work. However, it is essential to understand that powerful semidefinite programming solvers as SEDUMI o Mosek can deal with quadratic functions, as presented in this paper, and it is not necessary to address the non-quadratic problem.*

4. Numerical Results

To validate the convex nonlinear model (7) of the 3 DOF crane with respect to the nonlinear model given in [27], the parameters given in Table 2 are considered with initial conditions $x(0) = 0$, $l(0) = 0.22$ [m], $\theta(0) = 0$. A unit-input pulse it is considered for both actuators from 1 [s] $\leq u(t) \leq$ 2 [s], such as the responses shown in Figure 3 are obtained.

Table 2. Parameters of the 3 degrees of freedom (DOF) crane.

Parameter	Value	Units
g	9.81	m/s^2
m	0.50	kg
M_x	1.655	kg
M_l	0.50	kg
D_x	100	Ns/m
D_l	82	Ns/m

For the sake of simplicity and page limitation, only the comparison of the displacements are showed here. As can be observer from Figure 3, the responses of both systems, the nonlinear and the qLPV, are practically the same due to the fact that the convex model represents exactly the nonlinear system on the sectors limited by the weighting functions. Then, this convex model can be used to design the convex controller.

For the controller design, Theorem 1 is solved by minimizing γ such as the LMI (16) is feasible. For such purpose, the YALMIP [39] toolbox and the SEDUMI [40] solver have been used. The computed attenuation level is $\gamma = 0.0314$. Note that an attenuation level $\gamma < 1$ guarantees a good robust performance against noise and disturbances. The resulting P matrix is:

$$P = \begin{bmatrix} 29.2989 & 33.8755 & 35.2641 & -25.3974 & -112.0708 & -1.1582 \\ 33.8755 & 41.7475 & 46.7113 & -29.3192 & -139.3964 & -1.8173 \\ 35.2641 & 46.7113 & 165.1513 & 2.0182 & -168.5182 & -5.5944 \\ -25.3974 & -29.3192 & 2.0182 & 87.9937 & 86.5479 & -4.0631 \\ -112.0708 & -139.3964 & -168.5182 & 86.5479 & 478.4657 & 8.7512 \\ -1.1582 & -1.8173 & -5.5944 & -4.0631 & 8.7512 & 1.4297 \end{bmatrix}.$$

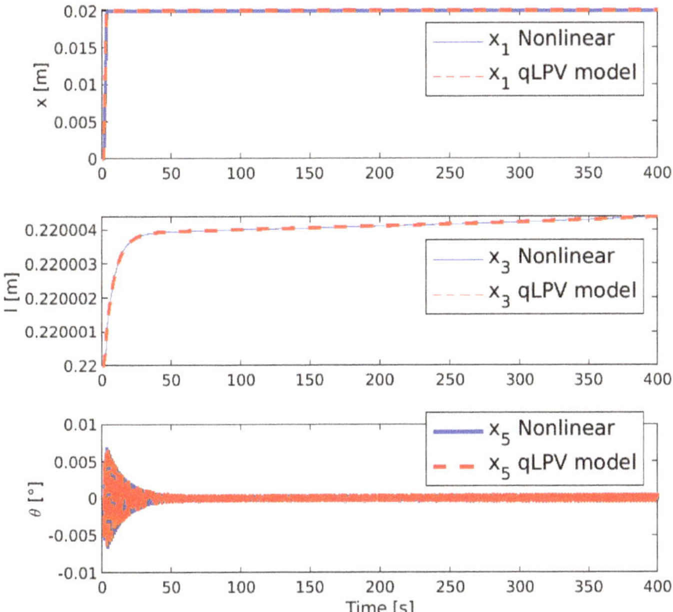

Figure 3. Comparison between the nonlinear and the quasi-Linear Parameter Varying (qLPV) models.

The numerical simulations of the controller were carried-out by considering the initial conditions given earlier and Gaussian random noise with zero mean and variance 0.2 in the measurements. The control objective is to displace the cart to track a reference consisting of a pulse oscillation of 1 [m] from the origin, for 20 [secs] in each position, and maintain the load in 0.4 [m] for $t \geq 5$ [s]. In addition, in order to evaluate the fault tolerance of the closed-loop system, an additive fault is applied to the input u_1, the fault is defined as follows:

$$\textbf{Fault } u_1 = \begin{cases} 0 & t < 35 \\ 15\% \text{ of } u_1 & 35 \leq t \leq 45 \\ 35\% \text{ of } u_1 & t > 45 \end{cases}.$$

The fault corresponds to a degradation of the force F_x given by the motor of the cart (25% and 35% of its nominal value). To include these faults, the matrix $G_h = B_h$, which represents the additive fault. The numerical simulations results are displayed in Figure 4.

As it can be observed, the controller tracks the demanded changes in the cart position and also maintains the load position. As a result that the proposed approach considers a passive fault-tolerant approach, the controller is robust to the actuator's fault. As it can be analyzed, the fault is compensated as soon as it appears, which means that its effect is practically eliminated from the system response.

Different fault scenarios were carried-out to test the controller performance and have been found that for actuator faults involving a degradation higher than 50%, the controller cannot reach the desired tracking position anymore. The effect of the noisy measured signals is reflected in the payload oscillation as this signal looks trembling. In addition, despite the continuous displacement, the load oscillation is attenuated maintaining a maximum of ± 2 degree with respect to the vertical. Nevertheless, this does not represent a limitation because the main objective was to maintain the desired position under partial faults. For more significant faults in magnitude, it would be necessary to replace the motor in order to guarantee the system's safety.

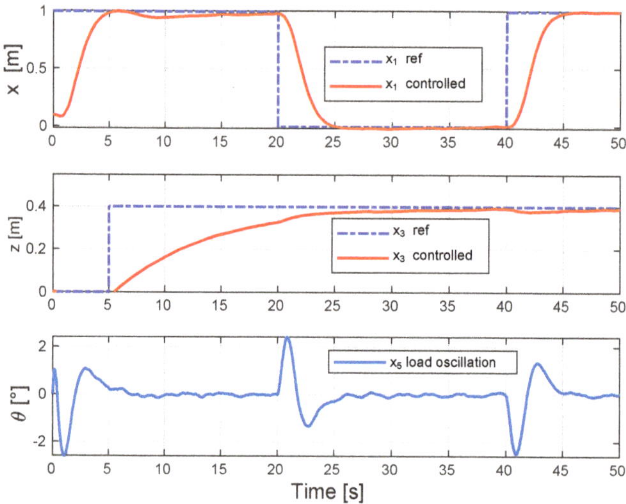

Figure 4. Control performance under an actuator fault.

5. Conclusions

This paper has presented a passive fault-tolerant controller for a 3 degree of freedom mechanical crane. First, a convex model of the 3 DOF crane has been proposed, representing the nonlinear dynamic by a set of linear models interpolated by nonlinear weighting functions. Then, a tracking fault-tolerant controller with $\mathcal{H}_\infty$ performance criteria has been developed over the nonlinear states' trajectories. The $\mathcal{H}_\infty$ performance guarantees robustness against measurement noise and partial faults. The numerical simulations results show the effectivity of the proposed method by tracking a predefined position of the cart and the load while the oscillations are attenuated despite the actuator faults. Future work will investigate the inclusion of measurement noise and will compare the development with a full nonlinear controller such as a nonlinear or sliding model controller.

Author Contributions: Conceptualization, F.-R.L.-E. and S.G.-P.; formal analysis, O.S.-E. and C.H.-G.; methodology, F.-R.L.-E. and G.V.-P.; software O.S.-E. and S.G.-P.; validation G.V.-P. and C.H.-G.; writing—original draft preparation, F.-R.L.-E. and G.V.-P.; writing—review and editing, S.G.-P. and O.S.-E. All authors have read and agreed to the published version of the manuscript.

Funding: The Consejo Estatal de Ciencia y Tecnológia del Estado de Chiapas financed this project under the grant number 1123. Aditional funding was provided from Conacyt, grants Projects 88 and 2759; and Tecnológico Nacional de México, grants 7641.20-P and 8017.20-P.

Conflicts of Interest: The authors declare no conflict of interest. The funders had no role in the design of the study; in the collection, analyses, or interpretation of data; in the writing of the manuscript, or in the decision to publish the results.

References

1. Blanke, M.; Kinnaert, M.; Lunze, J.; Staroswiecki, M.; Schröder, J. *Diagnosis and Fault-Tolerant Control*; Springer: Berlin/Heidelberg, Germany, 2006; Volume 691.
2. Martínez-García, C.; Puig, V.; Astorga-Zaragoza, C.M.; Madrigal-Espinosa, G.; Reyes-Reyes, J. Estimation of Actuator and System Faults Via an Unknown Input Interval Observer for Takagi–Sugeno Systems. *Processes* **2020**, *8*, 61. [CrossRef]
3. Witczak, M. Fault diagnosis and fault-tolerant control strategies for non-linear systems. *Lect. Notes Electr. Eng.* **2014**, *266*, 375–392.
4. García, C.M.; Puig, V.; Astorga-Zaragoza, C.; Osorio-Gordillo, G. Robust fault estimation based on interval takagi–sugeno unknown input observer. *IFAC-PapersOnLine* **2018**, *51*, 508–514. [CrossRef]
5. Nagy-Kiss, A.M.; Ichalal, D.; Schutz, G.; Ragot, J. Fault tolerant control for uncertain descriptor multi-models with application to wastewater treatment plant. In Proceedings of the American Control Conference (ACC), Chicago, IL, USA, 1–3 July 2015; pp. 5718–5725.
6. Wu, Y.; Dong, J. Fault detection for T–S fuzzy systems with partly unmeasurable premise variables. *Fuzzy Sets Syst.* **2018**, *338*, 136–156. [CrossRef]
7. Quintana, D.; Estrada-Manzo, V.; Bernal, M. Real-time parallel distributed compensation of an inverted pendulum via exact Takagi–Sugeno models. In Proceedings of the 2017 14th International Conference on Electrical Engineering, Computing Science and Automatic Control (CCE), Mexico City, Mexico, 20–22 October 2017; pp. 1–5.
8. López-Estrada, F.R.; Theilliol, D.; Astorga-Zaragoza, C.M.; Ponsart, J.C.; Valencia-Palomo, G.; Camas-Anzueto, J. Fault diagnosis observer for descriptor Takagi–Sugeno systems. *Neurocomputing* **2019**, *331*, 10–17. [CrossRef]
9. Gómez-Peñate, S.; López-Estrada, F.R.; Valencia-Palomo, G.; Rotondo, D.; Enríquez-Zárate, J. Actuator and sensor fault estimation based on a proportional-integral quasi-LPV observer with inexact scheduling parameters. *IFAC-PapersOnLine* **2019**, *52*, 100–105. [CrossRef]
10. Gómez-Peñate, S.; Valencia-Palomo, G.; López-Estrada, F.R.; Astorga-Zaragoza, C.M.; Osornio-Rios, R.A.; Santos-Ruiz, I. Sensor fault diagnosis based on a sliding mode and unknown input observer for Takagi–Sugeno systems with uncertain premise variables. *Asian J. Control* **2019**, *21*, 339–353. [CrossRef]
11. Chen, L.; Alwi, H.; Edwards, C. On the synthesis of an integrated active LPV FTC scheme using sliding modes. *Automatica* **2019**, *110*, 108536. [CrossRef]
12. Ohtake, H.; Tanaka, K.; Wang, H.O. Fuzzy modeling via sector nonlinearity concept. *Integr. Comput. Aided Eng.* **2003**, *10*, 333–341. [CrossRef]
13. Rotondo, D. *Advances in Gain-Scheduling and Fault Tolerant Control Techniques*; Springer: Berlin/Heidelberg, Germany, 2017.
14. López-Estrada, F.R.; Rotondo, D.; Valencia-Palomo, G. A Review of Convex Approaches for Control, Observation and Safety of Linear Parameter Varying and Takagi–Sugeno Systems. *Processes* **2019**, *7*, 814. [CrossRef]
15. Zhang, Y.; Jiang, J. Bibliographical review on reconfigurable fault-tolerant control systems. *Ann. Rev. Control* **2008**, *32*, 229–252. [CrossRef]
16. Yu, X.; Zhang, Y. Design of passive fault-tolerant flight controller against actuator failures. *Chin. J. Aeronaut.* **2015**, *28*, 180–190. [CrossRef]
17. Nasiri, A.; Nguang, S.K.; Swain, A.; Almakhles, D. Passive actuator fault tolerant control for a class of MIMO nonlinear systems with uncertainties. *Int. J. Control* **2019**, *92*, 693–704. [CrossRef]
18. Liu, Z.; Liu, J.; He, W. Robust adaptive fault tolerant control for a linear cascaded ODE-beam system. *Automatica* **2018**, *98*, 42–50. [CrossRef]
19. Nemati, F.; Hamami, S.M.S.; Zemouche, A. A nonlinear observer-based approach to fault detection, isolation and estimation for satellite formation flight application. *Automatica* **2019**, *107*, 474–482. [CrossRef]
20. Guzmán-Rabasa, J.A.; López-Estrada, F.R.; González-Contreras, B.M.; Valencia-Palomo, G.; Chadli, M.; Perez-Patricio, M. Actuator fault detection and isolation on a quadrotor unmanned aerial vehicle modeled as a linear parameter-varying system. *Meas. Control* **2019**, *52*, 1228–1239. [CrossRef]
21. Kim, Y.S.; Hong, K.S.; Sul, S.K. Anti-sway control of container cranes: Inclinometer, observer, and state feedback. *Int. J. Control Autom. Syst.* **2004**, *2*, 435–449.

22. Shi, K.; Wang, B.; Yang, L.; Jian, S.; Bi, J. Takagi–Sugeno fuzzy generalized predictive control for a class of nonlinear systems. *Nonlinear Dyn.* **2017**, *89*, 169–177. [CrossRef]
23. Almutairi, N.B.; Zribi, M. Sliding mode control of a three-dimensional overhead crane. *J. Vib. Control* **2009**, *15*, 1679–1730. [CrossRef]
24. Castillo, I.; Vázquez, C.; Fridman, L. Overhead crane control through LQ singular surface design MATLAB Toolbox. In Proceedings of the American Control Conference (ACC), Chicago, IL, USA, 1–3 July 2015; pp. 5847–5852.
25. Maghsoudi, M.J.; Mohamed, Z.; Sudin, S.; Buyamin, S.; Jaafar, H.; Ahmad, S. An improved input shaping design for an efficient sway control of a nonlinear 3D overhead crane with friction. *Mech. Syst. Signal Process.* **2017**, *92*, 364–378. [CrossRef]
26. Vu, N.T.T.; Thanh, P.T.; Duong, P.X.; Phuoc, N.D. Robust Adaptive Control of 3D Overhead Crane System. In *Adaptive Robust Control Systems*; IntechOpen: London, UK, 2017.
27. Abdullahi, A.M.; Mohamed, Z.; Selamat, H.; Pota, H.R.; Abidin, M.Z.; Ismail, F.; Haruna, A. Adaptive output-based command shaping for sway control of a 3D overhead crane with payload hoisting and wind disturbance. *Mech. Syst. Signal Process.* **2018**, *98*, 157–172. [CrossRef]
28. Chen, W.; Wu, Q.; Tafazzoli, E.; Saif, M. Actuator fault diagnosis using high-order sliding mode differentiator (HOSMD) and its application to a laboratory 3D crane. *IFAC Proc. Vol.* **2008**, *41*, 4809–4814. [CrossRef]
29. Tan, C.P.; Edwards, C. A robust sensor fault tolerant control scheme implemented on a crane. *Asian J. Control* **2007**, *9*, 340–344. [CrossRef]
30. Chen, W.; Saif, M. Actuator fault diagnosis for a class of nonlinear systems and its application to a laboratory 3D crane. *Automatica* **2011**, *47*, 1435–1442. [CrossRef]
31. Kiriakidis, K. Robust stabilization of the Takagi–Sugeno fuzzy model via bilinear matrix inequalities. *IEEE Trans. Fuzzy Syst.* **2001**, *9*, 269–277. [CrossRef]
32. Adeli, M.; Zarabadipour, H.; Zarabadi, S.H.; Shoorehdeli, M.A. Anti-swing control for a double-pendulum-type overhead crane via parallel distributed fuzzy LQR controller combined with genetic fuzzy rule set selection. In Proceedings of the IEEE International Conference on Control System, Computing and Engineering, Penang, Malaysia, 25–27 November 2011; pp. 306–311.
33. Hilhorst, G.; Pipeleers, G.; Michiels, W.; Oliveira, R.C.; Peres, P.L.D.; Swevers, J. Reduced-order $\mathcal{H}_2/\mathcal{H}_\infty$ control of discrete-time LPV systems with experimental validation on an overhead crane test setup. In Proceedings of the American Control Conference (ACC), Chicago, IL, USA, 1–3 July 2015; pp. 125–130.
34. Zhao, L.; Li, L. Robust stabilization of T–S fuzzy discrete systems with actuator saturation via PDC and non-PDC law. *Neurocomputing* **2015**, *168*, 418–426. [CrossRef]
35. Rabaoui, B.; Rodrigues, M.; Hamdi, H.; BenHadj Braiek, N. A model reference tracking based on an active fault tolerant control for LPV systems. *Int. J. Adapt. Control Signal Process.* **2018**, *32*, 839–857. [CrossRef]
36. Morato, M.M.; Sename, O.; Dugard, L. LPV-MPC Fault Tolerant Control of Automotive Suspension Dampers. *IFAC-PapersOnLine* **2018**, *51*, 31–36. [CrossRef]
37. Petrehuş, P.; Lendek, Z.; Raica, P. Fuzzy modeling and design for a 3D Crane. *IFAC Proc. Vol.* **2013**, *46*, 479–484. [CrossRef]
38. Márquez, R.; Guerra, T.M.; Bernal, M.; Kruszewski, A. Asymptotically necessary and sufficient conditions for Takagi–Sugeno models using generalized non-quadratic parameter-dependent controller design. *Fuzzy Sets Syst.* **2017**, *306*, 48–62. [CrossRef]
39. Lofberg, J. YALMIP: A Toolbox for modeling and optimization in MATLAB. In Proceedings of the 2004 IEEE International Conference on Robotics and Automation (IEEE Cat. No.04CH37508), New Orleans, LA, USA, 2–4 September 2004.
40. Sturm, J. Using SeDuMi 1.02, a MATLAB toolbox for optimization over symmetric cones. *Optim. Methods Softw.* **1999**, *11*, 625–653. [CrossRef]

© 2020 by the authors. Licensee MDPI, Basel, Switzerland. This article is an open access article distributed under the terms and conditions of the Creative Commons Attribution (CC BY) license (http://creativecommons.org/licenses/by/4.0/).

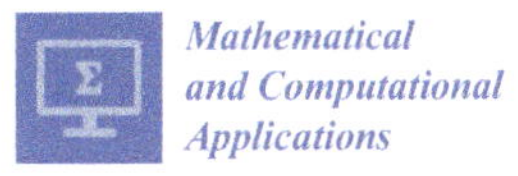

Article

Design of a Nonhomogeneous Nonlinear Synchronizer and Its Implementation in Reconfigurable Hardware

Jesus R. Pulido-Luna [1,†], Jorge A. López-Rentería [1,2,†] and Nohe R. Cazarez-Castro [1,*]

[1] Departamento de Ingeniería Eléctrica y Electrónica, Tecnológico Nacional de México–Instituto Tecnológico de Tijuana, Av Castillo de Chapultepec 562, Tomas Aquino, Tijuana 22414, B.C., Mexico; jesus.pulido19@tectijuana.edu.mx (J.R.P.-L.); jorge.lopez@tectijuana.edu.mx (J.A.L.-R.)

[2] Consejo Nacional de Ciencia y Tecnología, Av. Insurgentes Sur 1582, Col. Crédito Constructor, Alcaldía Benito Juárez, Mexico City 03940, C.P., Mexico

* Correspondence: nohe@ieee.org

† These authors contributed equally to this work.

Received: 19 July 2020; Accepted: 13 August 2020; Published: 14 August 2020

Abstract: In this work, a generalization of a synchronization methodology applied to a pair of chaotic systems with heterogeneous dynamics is given. The proposed control law is designed using the error state feedback and Lyapunov theory to guarantee asymptotic stability. The control law is used to synchronize two systems with different number of scrolls in their dynamics and defined in a different number of pieces. The proposed control law is implemented in an FPGA in order to test performance of the synchronization schemes.

Keywords: chaos; synchronization; FPGA; UDS

---

1. Introduction

Dynamical systems that exhibit chaotic behavior have proven to be very useful in science and engineering, for this same reason it is important to look for implementation alternatives that are fast and reliable. FPGAs are a very useful technology for these types of needs, due to their flexibility and the user friendly programming approach.

Since the discovery of systems with chaotic behavior, multiple analysis have been carried out [1–4], and the topic of synchronization of this class of systems has been a highly studied topic during the last 30 years [5]. This is due to the mistaken perception that this class of systems cannot be synchronized due to the complexity of their dynamics. This myth vanished in 1983 thanks to Yamada and Fujisaka [6] where a methodology for the synchronization of two chaotic systems using bidirectional coupling is presented, meanwhile in 1990 Pecora and Carroll [7] proposed the synchronization of the drive and response systems with different initial conditions. Since then, a wide series of alternative methodologies for the synchronization of chaotic systems have been developed [8–16] and thanks to this methodologies, a vast quantity of possible applications have been found in science and engineering, from physics [17,18], optics [19,20], biology [21–23], chemistry [24,25] and specially in the branch of secure communications [26–28].

A wide variety of chaotic systems have been implemented in circuits [29–33], this class of circuit implementations have certain disadvantages, such as the fact that they need very large changes in case the system wants to be modified. FPGAs have shown great flexibility in this regard [34–36] and although the original system changes, the only significant change is the reprogramming of the FPGA, which represents a great advantage when working on prototyping of new applications, a situation that represents cost savings and implementation times.

The aim of this work is to generalize a master–slave synchronization methodology in order to synchronize two chaotic systems with heterogeneous dynamics, which means that the master system and the slave system do not need to present the same behavior through time and it is not necessary that they are defined in the same number of parts, for example, the master system can be a system defined as a piecewise system, with n parts, while the slave system can be defined in a single part, with a single domain. In addition to this, the implementation of the most representative synchronization scheme is carried out in an FPGA, which allows these schemes to be used in multiple different applications.

The rest of this work is divided in the following way: in Section 2 the systems with which we will work are presented and a brief description of them is given; in Section 3, four different synchronization schemes are presented, including the methodology; in Section 4 the implementation of one of the schemes in an FPGA is presented and the results obtained are shown; finally, in Section 5 the conclusions are presented.

2. Preliminaries

This section presents in a non exhaustive way the dynamical systems that will be used in the rest of the paper.

2.1. The Generalized Lorenz System

Consider the generalized Lorenz system (GLS) defined in [37] as

$$\dot{x} = \begin{pmatrix} a_{11} & a_{12} & 0 \\ a_{21} & a_{22} & 0 \\ 0 & 0 & a_{33} \end{pmatrix} x + x_1 \begin{pmatrix} 0 & 0 & 0 \\ 0 & 0 & -1 \\ 0 & 1 & 0 \end{pmatrix} x, \tag{1}$$

where $x = (x_1, x_2, x_3)^T$. Four typical chaotic systems can be specified from (1): (*i*) Classical Lorenz system with $a_{12} = -a_{11} = a$, $a_{21} = c$, $a_{22} = -1$ and $a_{33} = -b$; (*ii*) Chen system with $a_{12} = -a_{11} = a$, $a_{21} = c - a$, $a_{22} = c$ and $a_{33} = -b$; (*iii*) Lü system using $a_{12} = -a_{11} = a$, $a_{21} = 0$, $a_{22} = c$ and $a_{33} = -b$; (*iv*) Unified chaotic system with $a_{12} = -a_{11} = 25 + \eta$, $a_{21} = 28 - 35\eta$, $a_{22} = 29\eta - 1$ and $a_{33} = -\frac{8+\eta}{3}$, where $a, b, c \in \mathbb{R}^+$ and $\eta \in [0, 1]$.

2.2. Unstable Dissipative Systems

Now consider a unstable dissipative system (UDS) defined in [38] as

$$\dot{\chi} = A\chi + B, \tag{2}$$

where $\chi = (\chi_1, \chi_2, \chi_3)^T$, $A = (\alpha_{ij})_{j=1}^3$ and B contains the switching law of the form

$$B = \begin{cases} B_1 & if \quad \chi \in D_1, \\ B_2 & if \quad \chi \in D_2, \\ \vdots & \vdots \qquad \vdots \\ B_k & if \quad \chi \in D_k, \end{cases} \tag{3}$$

with $B_k = (b_{k1}, b_{k2}, b_{k3})^T$. It is possible to define two types of UDS, and two types of correspond equilibria.

Definition 1 (Campos-Cantón et al. [39]). *A system given by (2) with eigenvalues λ_i, $i = 1, 2, 3$, satisfying $\sum_{i=1}^3 \lambda_i < 0$. Then, the system is said to be:*

(*i*) *An UDS Type I, if one eigenvalue is negative real and the other two are complex conjugate with a positive real part.*

(*ii*) *An UDS Type II, if one eigenvalue is positive real and the other two are complex conjugate with a negative real part.*

For the equilibria, their two types are defined accordingly. In Definition 1, item (i) implies that the UDS Type I is dissipative in one of its components but oscillatory unstable in the other two, while item (ii) implies that an UDS Type II is dissipative and oscillatory in two of their components but unstable in the other one. Some chaotic dynamical systems may relate to these two types of UDS around equlibria, systems as the ones in [29,40–42] can be characterized through a combination of UDS Type I and Type II.

3. Synchronization Scheme

The synchronization scheme diagram is depicted the Figure 1. The output X of the master FPGA is the input of the slave FPGA, the controller takes this input X and the output Y of the slave system inside of the slave FPGA and compensates the slave system output, which means that $\lim_{t\to\infty} |Y - X| = 0$.

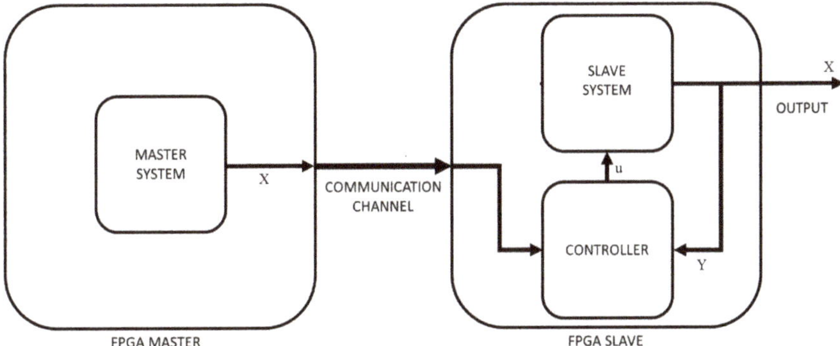

Figure 1. Synchronization scheme diagram.

The rest of this Section presents four master–slave synchronization schemes.

3.1. Master UDS–Slave GLS

Consider the master system as

$$\dot{\chi} = \begin{pmatrix} \alpha_{11} & \alpha_{12} & \alpha_{13} \\ \alpha_{21} & \alpha_{22} & \alpha_{23} \\ \alpha_{31} & \alpha_{32} & \alpha_{33} \end{pmatrix} \chi + \begin{pmatrix} b_{k_1} \\ b_{k2} \\ b_{k3} \end{pmatrix}, \tag{4}$$

while the slave system is defined as

$$\dot{x} = \begin{pmatrix} a_{11} & a_{12} & 0 \\ a_{21} & a_{22} & 0 \\ 0 & 0 & a_{33} \end{pmatrix} x + x_1 \begin{pmatrix} 0 & 0 & 0 \\ 0 & 0 & -1 \\ 0 & 1 & 0 \end{pmatrix} x + u, \tag{5}$$

where $u = (u_1, u_2, u_3)^T$. The error vector is defined as $e = x - \chi$, and is possible to obtain

$$\dot{e} = \begin{pmatrix} a_{11} & a_{12} & 0 \\ a_{21} & a_{22} & 0 \\ 0 & 0 & a_{33} \end{pmatrix} x + x_1 \begin{pmatrix} 0 & 0 & 0 \\ 0 & 0 & -1 \\ 0 & 1 & 0 \end{pmatrix} x - \begin{pmatrix} \alpha_{11} & \alpha_{12} & \alpha_{13} \\ \alpha_{21} & \alpha_{22} & \alpha_{23} \\ \alpha_{31} & \alpha_{32} & \alpha_{33} \end{pmatrix} \chi - \begin{pmatrix} b_{k1} \\ b_{k2} \\ b_{k3} \end{pmatrix} + u. \tag{6}$$

In order to stabilize the error system, the proposed Lyapunov function is

$$V = \frac{1}{2}\left(e_1^2 + e_2^2 + e_3^2\right), \tag{7}$$

whose derivative is

$$\begin{aligned}\dot{V} = \; & e_1\left(a_{11}x_1 + a_{12}x_2 - \alpha_{11}\chi_1 - \alpha_{12}\chi_2 - \alpha_{13}\chi_3 - b_{k1} + u_1\right) + e_2\left(a_{21}x_1 + a_{22}x_2 - x_1x_3\right. \\ & -\alpha_{21}\chi_1 - \alpha_{22}\chi_2 - \alpha_{23}\chi_3 - b_{k2} + u_2) + e_3\left(a_{33}x_3 + x_1x_2 - \alpha_{31}\chi_1 - \alpha_{32}\chi_2 - \alpha_{33}\chi_3\right. \\ & \left. -b_{k3} + u_3\right),\end{aligned} \tag{8}$$

which allows to design the control law u as

$$u = -Pe - \begin{pmatrix}\alpha_{11} & \alpha_{12} & \alpha_{13} \\ \alpha_{21} & \alpha_{22} & \alpha_{23} \\ \alpha_{31} & \alpha_{32} & \alpha_{33}\end{pmatrix}\chi - \begin{pmatrix}b_{k1} \\ b_{k2} \\ b_{k3}\end{pmatrix} + \begin{pmatrix}a_{11} & a_{12} & 0 \\ a_{21} & a_{22} & 0 \\ 0 & 0 & a_{33}\end{pmatrix}x + x_1\begin{pmatrix}0 & 0 & 0 \\ 0 & 0 & -1 \\ 0 & 1 & 0\end{pmatrix}x, \tag{9}$$

where $e = (e_1, e_2, e_3)^T$ is the error vector, and $P = P^T = \text{diag}\{p_1^2, p_2^2, p_3^2\}$ is a diagonal matrix of parameters selected to ensure negativeness of (8).

3.2. Master GLS–Slave UDS

For this scheme the master system is considered as

$$\dot{x} = \begin{pmatrix}a_{11} & a_{12} & 0 \\ a_{12} & a_{22} & 0 \\ 0 & 0 & a_{33}\end{pmatrix}x + x_1\begin{pmatrix}0 & 0 & 0 \\ 0 & 0 & -1 \\ 0 & 1 & 0\end{pmatrix}x, \tag{10}$$

while the slave system is

$$\dot{\chi} = \begin{pmatrix}\alpha_{11} & \alpha_{12} & \alpha_{13} \\ \alpha_{21} & \alpha_{22} & \alpha_{23} \\ \alpha_{31} & \alpha_{32} & \alpha_{33}\end{pmatrix}\chi + \begin{pmatrix}b_{k1} \\ b_{k2} \\ b_{k3}\end{pmatrix} + u. \tag{11}$$

The error vector is defined as $e = \chi - x$, consequently the error dynamics are represented by

$$\dot{e} = \begin{pmatrix}\alpha_{11} & \alpha_{12} & \alpha_{13} \\ \alpha_{21} & \alpha_{22} & \alpha_{23} \\ \alpha_{31} & \alpha_{32} & \alpha_{33}\end{pmatrix}\chi + \begin{pmatrix}b_{k1} \\ b_{k2} \\ b_{k3}\end{pmatrix} - \begin{pmatrix}a_{11} & a_{12} & 0 \\ a_{12} & a_{22} & 0 \\ 0 & 0 & a_{33}\end{pmatrix}x - x_1\begin{pmatrix}0 & 0 & 0 \\ 0 & 0 & -1 \\ 0 & 1 & 0\end{pmatrix}x + u. \tag{12}$$

Similarly to the previous scheme, the proposed Lyapunov function is

$$V = \frac{1}{2}\left(e_1^2 + e_2^2 + e_3^2\right), \tag{13}$$

while in this scheme the derivative results in

$$\begin{aligned}\dot{V} = \; & e_1\left(\alpha_{11}\chi_1 + \alpha_{12}\chi_2 + \alpha_{13}\chi_3 + b_{k1} - a_{11}x_1 - a_{12}x_2 + u_1\right) + e_2\left(\alpha_{21}\chi_1 + \alpha_{22}\chi_2 + \alpha_{23}\chi_3\right. \\ & -a_{21}x_1 - a_{22}x_2 + x_1x_3 + b_{k2} + u_2) + e_3\left(\alpha_{31}\chi_1 + \alpha_{32}\chi_2 + \alpha_{33}\chi_3 - a_{33}x_3 - x_1x_2 + b_{k3}\right. \\ & \left. +u_3\right).\end{aligned} \tag{14}$$

Under the before considerations, the proposed control law is

$$u = -Pe - \begin{pmatrix}\alpha_{11} & \alpha_{12} & \alpha_{13} \\ \alpha_{21} & \alpha_{22} & \alpha_{23} \\ \alpha_{31} & \alpha_{32} & \alpha_{33}\end{pmatrix}\chi - \begin{pmatrix}b_{k1} \\ b_{k2} \\ b_{k3}\end{pmatrix} + \begin{pmatrix}a_{11} & a_{12} & 0 \\ a_{12} & a_{22} & 0 \\ 0 & 0 & a_{33}\end{pmatrix}x + x_1\begin{pmatrix}0 & 0 & 0 \\ 0 & 0 & -1 \\ 0 & 1 & 0\end{pmatrix}x, \tag{15}$$

where, in the same form as before $P = P^T = \text{diag}\{p_1^2, p_2^2, p_3^2\}$ is a matrix of parameters selected to ensure the negativeness of (14).

3.3. Master GLS–Slave GLS

The master system for this synchronization scheme is

$$\dot{x} = \begin{pmatrix} a_{11} & a_{12} & 0 \\ a_{21} & a_{22} & 0 \\ 0 & 0 & a_{33} \end{pmatrix} x + x_1 \begin{pmatrix} 0 & 0 & 0 \\ 0 & 0 & -1 \\ 0 & 1 & 0 \end{pmatrix} x, \tag{16}$$

However, in this scheme the slave system is defined as

$$\dot{y} = \begin{pmatrix} b_{11} & b_{12} & 0 \\ b_{21} & b_{22} & 0 \\ 0 & 0 & b_{33} \end{pmatrix} y + y_1 \begin{pmatrix} 0 & 0 & 0 \\ 0 & 0 & -1 \\ 0 & 1 & 0 \end{pmatrix} y + u, \tag{17}$$

Once again, $u = (u_1, u_2, u_3)^T$ is the controller. The error is defined as $e = y - x$, and its dynamics is given by

$$\dot{e} = \begin{pmatrix} b_{11} & b_{12} & 0 \\ b_{21} & b_{22} & 0 \\ 0 & 0 & b_{33} \end{pmatrix} y - \begin{pmatrix} a_{11} & a_{12} & 0 \\ a_{21} & a_{22} & 0 \\ 0 & 0 & a_{33} \end{pmatrix} x + \begin{pmatrix} 0 & 0 & 0 \\ 0 & 0 & -1 \\ 0 & 1 & 0 \end{pmatrix} (y_1 y - x_1 x) + u. \tag{18}$$

In order to stabilize the error dynamics, the proposed Lyapunov function is

$$V = \frac{1}{2}\left(e_1^2 + e_2^2 + e_3^2\right), \tag{19}$$

where the derivative is

$$\begin{aligned} \dot{V} = \; & e_1 \left(b_{11}y_1 + b_{12}y_2 - a_{11}x_1 - a_{12}x_2 + u_1\right) + e_2 \left(b_{21}y_1 + b_{22}y_2 - a_{21}x_1 - a_{22}x_2 - y_1y_3\right. \\ & \left. + x_1x_3 + u_2\right) + e_3 \left(b_{33}y_3 - a_{33}x_3 + y_1y_2 - x_1x_2 + u_3\right), \end{aligned} \tag{20}$$

which allows to design the controll law u as

$$u = -Pe - \begin{pmatrix} b_{11} & b_{12} & 0 \\ b_{21} & b_{22} & 0 \\ 0 & 0 & b_{33} \end{pmatrix} y + \begin{pmatrix} a_{11} & a_{12} & 0 \\ a_{21} & a_{22} & 0 \\ 0 & 0 & a_{33} \end{pmatrix} x - \begin{pmatrix} 0 & 0 & 0 \\ 0 & 0 & -1 \\ 0 & 1 & 0 \end{pmatrix} (y_1 y - x_1 x). \tag{21}$$

3.4. Master UDS–Slave UDS

Finally, the master system for this scheme is

$$\dot{\chi} = \begin{pmatrix} \alpha_{11} & \alpha_{12} & \alpha_{13} \\ \alpha_{21} & \alpha_{22} & \alpha_{23} \\ \alpha_{31} & \alpha_{32} & \alpha_{33} \end{pmatrix} \chi + \begin{pmatrix} b_{k1} \\ b_{k2} \\ b_{k3} \end{pmatrix}, \tag{22}$$

and the slave system is of the form

$$\dot{\varphi} = \begin{pmatrix} \beta_{11} & \beta_{12} & \beta_{13} \\ \beta_{21} & \beta_{22} & \beta_{23} \\ \beta_{31} & \beta_{32} & \beta_{33} \end{pmatrix} \varphi + \begin{pmatrix} c_{k1} \\ c_{k2} \\ c_{k3} \end{pmatrix} + u. \tag{23}$$

The dynamics of the error $\varphi - \chi$ is given by

$$\dot{e} = \begin{pmatrix} \beta_{11} & \beta_{12} & \beta_{13} \\ \beta_{21} & \beta_{22} & \beta_{23} \\ \beta_{31} & \beta_{32} & \beta_{33} \end{pmatrix} \varphi - \begin{pmatrix} \alpha_{11} & \alpha_{12} & \alpha_{13} \\ \alpha_{21} & \alpha_{22} & \alpha_{23} \\ \alpha_{31} & \alpha_{32} & \alpha_{33} \end{pmatrix} \chi + \begin{pmatrix} c_{k1} \\ c_{k2} \\ c_{k3} \end{pmatrix} - \begin{pmatrix} b_{k1} \\ b_{k2} \\ b_{k3} \end{pmatrix} + u, \tag{24}$$

and once again, the proposed Lyapunov function is

$$V = \frac{1}{2}\left(e_1^2 + e_2^2 + e_3^2\right), \tag{25}$$

with derivative

$$\begin{aligned} \dot{V} = \; & e_1\left(\beta_{11}\varphi_1 + \beta_{12}\varphi_2 + \beta_{13}\varphi_3 - \alpha_{11}\chi_1 - \alpha_{12}\chi_2 - \alpha_{13}\chi_3 + c_{k1} - b_{k1} + u_1\right) + e_2\left(\beta_{21}\varphi_1\right. \\ & \left. + \beta_{22}\varphi_2 + \beta_{23}\varphi_3 - \alpha_{21}\chi_1 - \alpha_{22}\chi_2 - \alpha_{23}\chi_3 + c_{k2} - b_{k2} + u_2\right) + e_3\left(\beta_{31}\varphi_1 + \beta_{32}\varphi_2\right. \\ & \left. + \beta_{33}\varphi_3 - \alpha_{31}\chi_1 - \alpha_{32}\chi_2 - \alpha_{33}\chi_3 + c_{k3} - b_{k3} + u_3\right). \end{aligned} \tag{26}$$

Consequently, the proposed control law is

$$u = -Pe - \begin{pmatrix} \beta_{11} & \beta_{12} & \beta_{13} \\ \beta_{21} & \beta_{22} & \beta_{23} \\ \beta_{31} & \beta_{32} & \beta_{33} \end{pmatrix} \varphi + \begin{pmatrix} \alpha_{11} & \alpha_{12} & \alpha_{13} \\ \alpha_{21} & \alpha_{22} & \alpha_{23} \\ \alpha_{31} & \alpha_{32} & \alpha_{33} \end{pmatrix} \chi - \begin{pmatrix} c_{k1} \\ c_{k2} \\ c_{k3} \end{pmatrix} + \begin{pmatrix} b_{k1} \\ b_{k2} \\ b_{k3} \end{pmatrix}. \tag{27}$$

4. Results

For the implementation of the synchronization scheme, the selected piecewise UDS given in [43] presents a four scrolls attractor and it is defined as

$$\dot{\chi} = \begin{pmatrix} 0 & 1 & 0 \\ 0 & 0 & 1 \\ -35.139 & -8.23 & -3.7 \end{pmatrix} \chi + \begin{pmatrix} 0 \\ 0 \\ B_k \end{pmatrix}, \tag{28}$$

where B_k is given by

$$B_k = \begin{cases} 21.0834 & \text{if} \quad \chi_1 > 0.5, \\ 14.055 & \text{if} \quad 0.3 < \chi_1 \leq 0.5, \\ 7.0278 & \text{if} \quad 0.1 < \chi_1 \leq 0.3, \\ 0 & \text{if} \quad \chi_1 \leq 0.1, \end{cases} \tag{29}$$

and the slave system is

$$\dot{x} = \begin{pmatrix} -16 & 16 & 0 \\ 45.6 & -1 & 0 \\ 0 & 0 & -4 \end{pmatrix} x + x_1 \begin{pmatrix} 0 & 0 & 0 \\ 0 & 0 & -1 \\ 0 & 1 & 0 \end{pmatrix} x + u. \tag{30}$$

Using the synchronization scheme described in the Section 3.1, the control law is defined as

$$u = -Pe - \begin{pmatrix} 0 & 1 & 0 \\ 0 & 0 & 1 \\ -35.139 & -8.23 & -3.7 \end{pmatrix} \chi - \begin{pmatrix} 0 \\ 0 \\ B_k \end{pmatrix} + \begin{pmatrix} -16 & 16 & 0 \\ 45.6 & -1 & 0 \\ 0 & 0 & -4 \end{pmatrix} x + x_1 \begin{pmatrix} 0 & 0 & 0 \\ 0 & 0 & -1 \\ 0 & 1 & 0 \end{pmatrix} x, \tag{31}$$

which will be implemented in a SPARTAN–3AN FPGA Starter Kit board from Xilinx, the general scheme can be appreciated in the Figure 2. It can be seen that in (**a**) the SPARTAN–3AN starter kit board in which the master system is implemented; in (**b**) the slave system is implemented with the control law (31); in (**c**) the Digital to Analog Converter (DAC) can be observed and in (**d**) the output data is acquired.

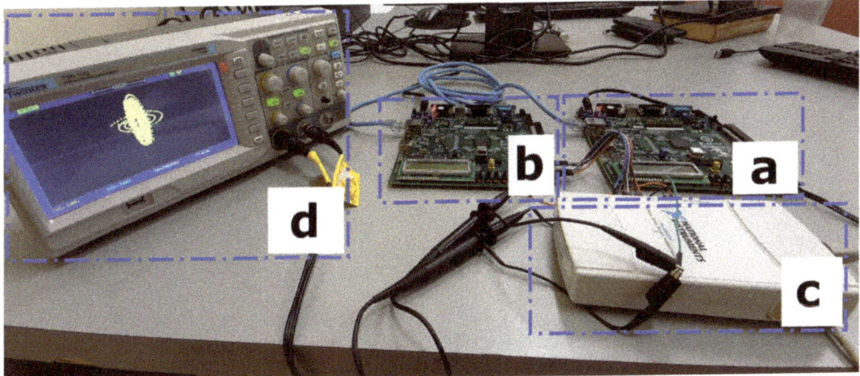

Figure 2. Synchronization scheme using two SPARTAN–3AN starter kit boards: (**a**) Master system. (**b**) Slave system. (**c**) DAC. (**d**) Data acquisition.

In order to implement the synchronization scheme in the SPARTAN–3AN starter kit, the Simulink® toolbox: `Xilinx System Generator` was used. In the Figure 3 can be observed the classical Lorenz system without any kind of control, implemented in Simulink® using this toolbox as example.

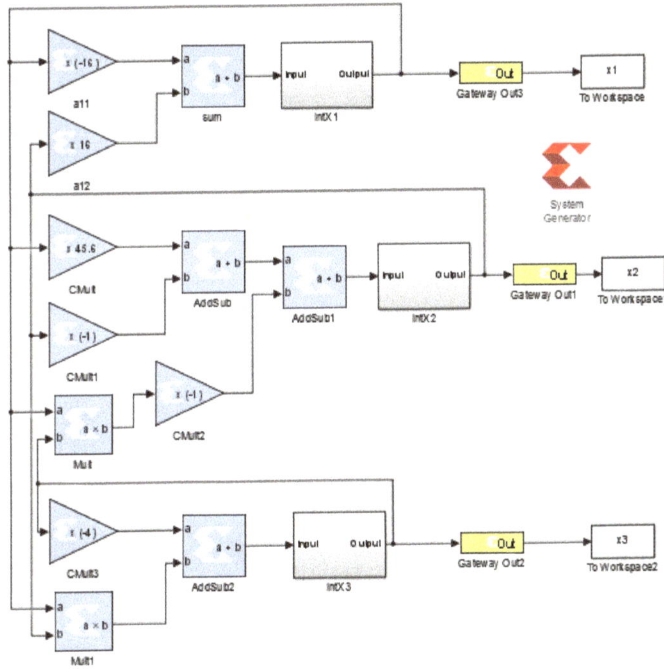

Figure 3. Classical Lorenz system implemented in Simulink® using the `Xilinx System Generator` toolbox.

When implementing the master system in the FPGA, it is possible to acquire the output signal using the DAC from the National Instruments module NI–6211. With this is possible to obtain the plane projections (χ_1, χ_2) and $\chi_{1,2}$ vs t, this can be appreciated in the Figure 4. This is also applicable to the projection on the plane (χ_1, χ_3), as can be seen in the Figure 5. The projections (χ_2, χ_3) are shown in the Figure 6, and the FPGA resources utilized by the master system is presented in Table 1.

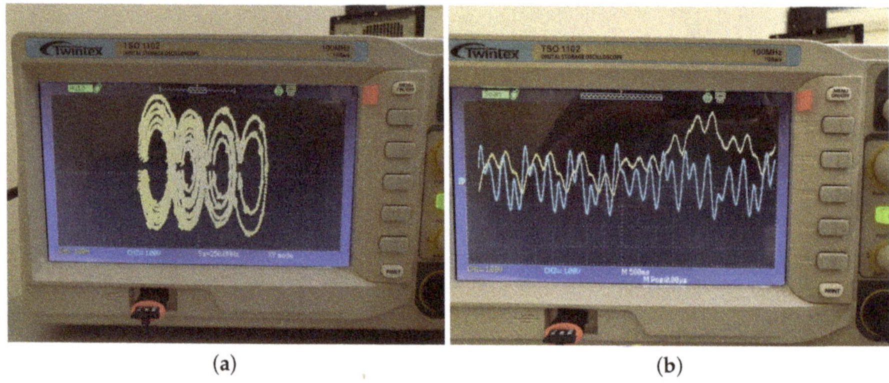

(**a**) (**b**)

Figure 4. Acquired data from the master system: (**a**) Projection on the plane (χ_1, χ_2). (**b**) $\chi_{1,2}$ vs t.

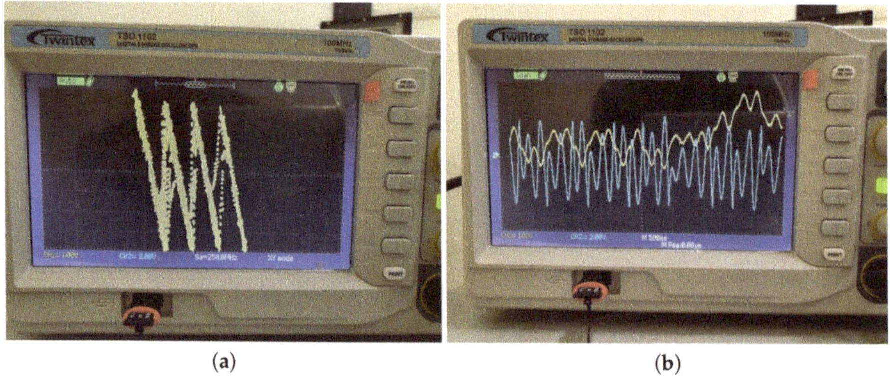

(**a**) (**b**)

Figure 5. Acquired data from the master system: (**a**) Projection on the plane (χ_1, χ_3). (**b**) $\chi_{1,3}$ vs t.

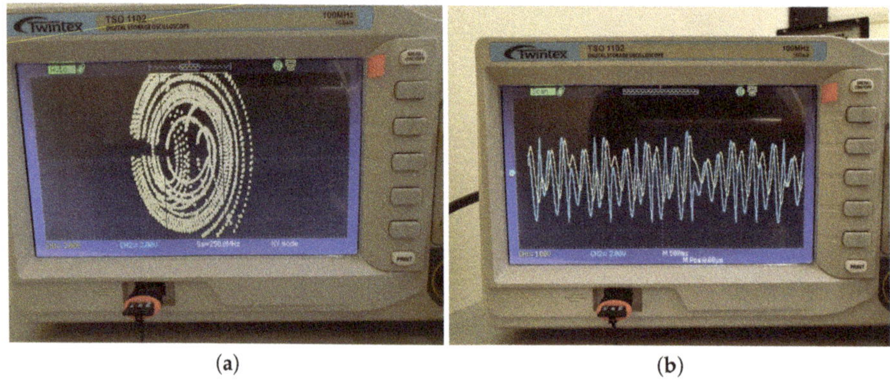

(**a**) (**b**)

Figure 6. Acquired data from the master system: (**a**) Projection on the plane (χ_2, χ_3). (**b**) $\chi_{3,3}$ vs t.

Table 1. FPGA utilization summary for the master 4–scrolls UDS.

Resources	Used	Available	Utilization
Number of Slice Flip Flops	159	11,776	1%
Number of 4 input LUTs	4177	11,776	35%
Number of occupied Slices	2473	5888	42%
Total Number of 4 input LUTs	4467	11,776	37%
Number of bounded IOBs	145	372	38%
Number of BUFGMUXs	1	24	4%

For the uncontrolled slave system, both of the projections on the planes (x_1, x_2) and (x_1, x_3) were obtained and they can be seen in Figure 7. The projections on the plane (x_2, x_3) are depicted in Figure 8. It is important to highlight that the slave system arises a two-scroll attractor while the master system presents four scrolls. Consequently, once implemented the control law, the slave system will change its dynamics undergoing a four-scroll attractor behavior with the shape of the master systems's phase portrait. The FPGA utilization summary for the slave system is presented in Table 2.

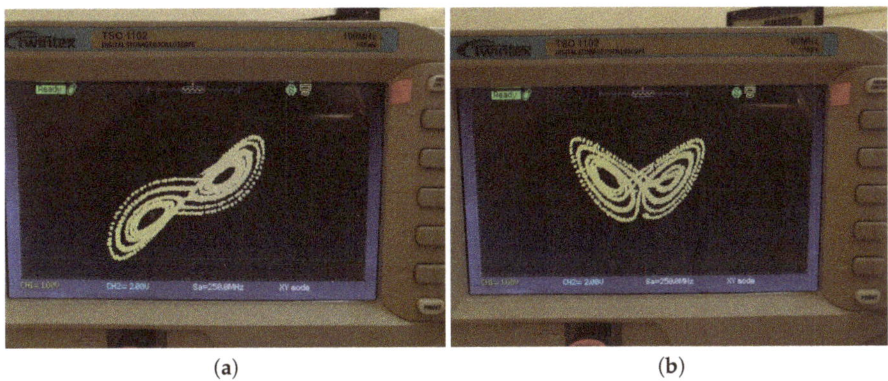

(a) **(b)**

Figure 7. Acquired data from the uncontrolled slave system: (**a**) Projection on the plane (x_1, x_2). (**b**) Projection on the plane (x_1, x_3).

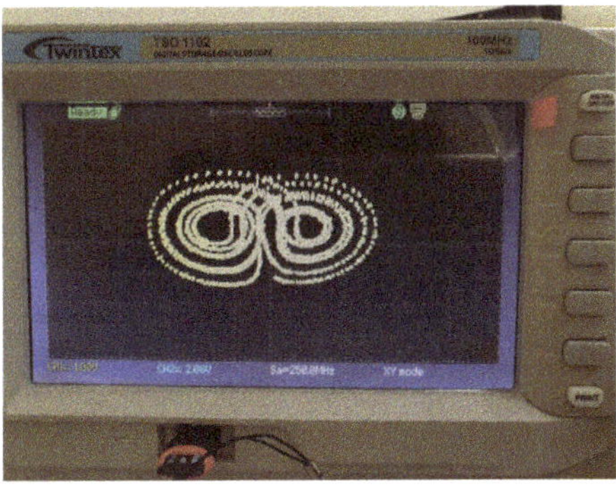

Figure 8. Projection on the plane (x_1, x_2) for the uncontrolled slave system.

Table 2. FPGA utilization summary for the uncontrolled Lorenz slave system.

Resources	Used	Available	Utilization
Number of Slice Flip Flops	170	11,776	1%
Number of 4 input LUTs	2373	11,776	20%
Number of occupied Slices	1491	5888	25%
Total Number of 4 input LUTs	2677	11,776	22%
Number of bounded IOBs	53	372	14%
Number of BUFGMUXs	1	24	4%
Number of MULT18X18SIOs	18	20	90%

For the synchronized slave system, both of the projections on the planes (x_1, x_2) and (x_1, x_3) were obtained and they can be observed in the Figure 9. The projections on the plane (x_2, x_3) are depicted in the Figure 10. It is easy to see that the two-scroll slave system in fact adopted the master system behavior presenting four scrolls. The resource utilization of the FPGA is shown in the Table 3, the utilization increases compared with the master and the uncontrolled slave, this is due to the integration of the controller u. This proves that in a low cost, entry-level FPGA like the SPARTAN–3AN the system are possible to implement by taking in consideration that the utilization of the FPGA exceeds the 60% of the resources. For the MULTI18X18IOs of the system the utilization is 100%, this problem can be avoided by implementing the synchronization scheme in a more powerful FPGA.

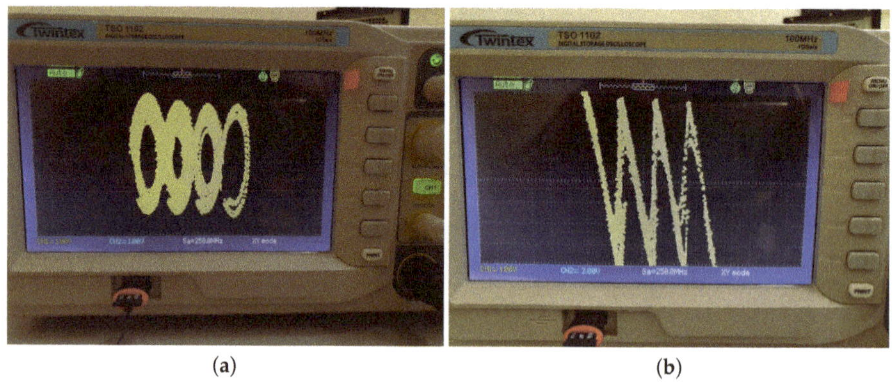

(a) (b)

Figure 9. Acquired data from the slave system: (**a**) Projection on the plane (x_1, x_2). (**b**) Projection on the plane (x_1, x_3).

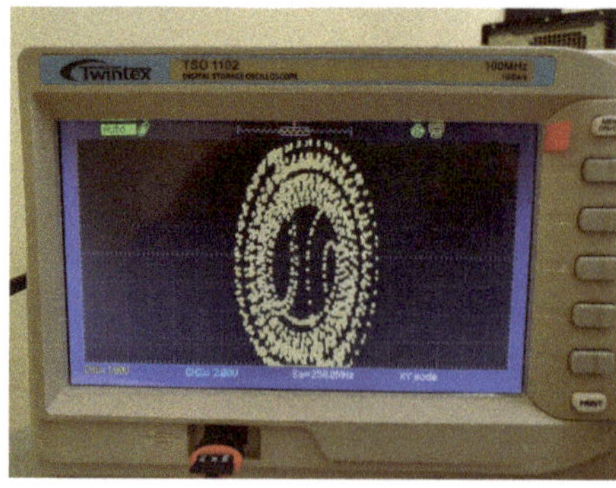

Figure 10. Projection on the plane (x_1, x_2) for the slave system.

Table 3. FPGA utilization summary for the controlled Lorenz slave system.

Resources	Used	Available	Utilization
Number of Slice Flip Flops	332	11,776	3%
Number of 4 input LUTs	5550	11,776	56%
Number of occupied Slices	2964	5888	67%
Total Number of 4 input LUTs	7244	11,776	62%
Number of bounded IOBs	194	372	52%
Number of BUFGMUXs	2	24	8%
Number of MULT18X18SIOs	20	20	100%

5. Conclusions

A synchronization scheme for systems with heterogeneous chaotic behavior was implemented in an FPGA, this synchronization scheme can synchronize a pair of chaotic systems defined by a different number of pieces, is possible to apply the scheme to UDSs which is a kind of generalization for synchronization of piecewise systems with different quantities of pieces on the master and slave systems, which allows the use in a wide variety of applications in science and engineering. The synchronization scheme gives to the slave system a the dynamics of the master system. The synchronization scheme is designed taking into account the error system between the master and the slave, adding a parameter matrix P that controls the synchronization speed, whenever the parameter is adequate. The controller guarantees a fast synchronization with a minimum error.

Author Contributions: Conceptualization, J.R.P.-L., J.A.L.-R. and N.R.C.-C.; methodology, J.R.P.-L., J.A.L.-R. and N.R.C.-C.; validation, J.R.P.-L., J.A.L.-R. and N.R.C.-C.; writing–original draft J.R.P.-L., J.A.L.-R. and N.R.C.-C.; Writing–review and editing, J.R.P.-L., J.A.L.-R. and N.R.C.-C. All authors have read and agreed to the published version of the manuscript.

Funding: This research was funded by CONACyT grant number A1-S-32341 and TecNM grant number 8085.20-P.

Acknowledgments: J.R.P.-L. would like to thank CONACyT for the master's degree scholarship.

Conflicts of Interest: The authors declare no conflict of interest. The funders had no role in the design of the study; in the collection, analyses, or interpretation of data; in the writing of the manuscript, or in the decision to publish the results.

Abbreviations

The following abbreviations are used in this manuscript:

FPGA	Field-Programmable Gate Array
GLS	Generalized Lorenz System
UDS	Unstable Dissipative System
DAC	Digital to Analogue Converter

References

1. Chua, L.; Komuro, M.; Matsumoto, T. The double scroll family. *IEEE Trans. Circuits Syst.* **1986**, *33*, 1073–1118. [CrossRef]
2. Mees, A.; Chapman, P. Homoclinic and heteroclinic orbits in the double scroll attractor. *IEEE Trans. Circuits Syst.* **1987**, *34*, 1115–1120. [CrossRef]
3. Devaney, R. *An Introduction to Chaotic Dynamical Systems*, 1st ed.; CRC Press: Boca Raton, FL, USA, 1989.
4. Silva, C. Shilnikov's theorem—A tutorial. *IEEE Trans. Circuits Syst.* **1993**, *64*, 675–682. [CrossRef]
5. Pikovsky, A.; Rosenblum, M.; Kurths, J. *Synchronization: A Universal Concept in Nonlinear Sciences*, 1st ed.; Cambridge University Press: Cambridge, UK, 2001.
6. Fujisaka, H.; Yamada, T. Stability Theory of Synchronized Motions in Coupled-Oscillator Systems. *Prog. Theor. Phys.* **1983**, *1983*, 32–47. [CrossRef]
7. Pecora, L.M.; Carroll, T.L. Synchronization in chaotic systems. *Phys. Rev. Lett.* **1990**, *64*, 821–824. [CrossRef]

8. Agiza, H.; Yassen, M. Synchronization of Rossler and Chen chaotic dynamical systems using active control. *Phys. Lett. A* **2001**, *278*, 191–197. [CrossRef]
9. Yassen, M. Chaos synchronization between two different chaotic systems using active control. *Chaos Solit. Fractals* **2005**, *23*, 131–140. [CrossRef]
10. Park, J. Chaos synchronization between two different chaotic dynamical systems. *Chaos Solit. Fractals* **2006**, *27*, 549–554. [CrossRef]
11. Mahmoud, G.; Bountis, T.; Mahmoud, E. Active control and global synchronization of the complex Chen and Lü systems. *Int. J. Bifurc. Chaos* **2007**, *17*, 4295–4308. [CrossRef]
12. Wu, X.; Chen, G.; Cai, J. Chaos synchronization of the master–slave generalized Lorenz system via linear state error feedback control. *Physica D* **2007**, *229*, 52–80. [CrossRef]
13. Zhang, T.; Feng, G. Output tracking of piecewise–linear systems via error feedback regulator with application to synchronization of nonlinear Chua's circuit. *IEEE Trans. Circuits Syst.* **2007**, *54*, 1852–1863. [CrossRef]
14. Oancea, S.; Grosu, F.; Lazar, A.; Grosu, I. Master–slave synchronization of Lorenz systems using a single controller. *Chaos Solit. Fractals* **2009**, *41*, 2575–2580. [CrossRef]
15. Mkaouar, H.; Boubaker, O. Chaos synchronization for master–slave piecewise linear systems: Application to Chua's circuit. *Commun. Nonlinear Sci. Numer. Simul.* **2012**, *17*, 1292–1302. [CrossRef]
16. Pérez, J.H.; Figueroa, M.; López, A. Synchronization of chaotic Akgul system by means of feedback linearization and pole placement. *IEEE Lat. Am. Trans.* **2017**, *15*, 249–256. [CrossRef]
17. Idowu, B.A.; Vincent, U.E. Synchronization and stabilization of chaotic dynamics in quasi–1D Bose–Einstein condensate. *J. Chaos* **2013**, *2013*, 1–8. [CrossRef]
18. Vaidyanathan, S. Anti–synchronization of rikitake two–disk dynamo chaotic systems via adaptive control method. *Int. J. ChemTech Res.* **2015**, *8*, 393–405.
19. Yuan, G.; Zhang, X.; Wuang, Z. Generation and synchronization of feedback–induced chaos in semiconductor ring lasers by injection locking. *Optik* **2014**, *125*, 1950–1953. [CrossRef]
20. Wuang, L.; Wu, Z.; Wu, J.; Xia, G. Long–haul dual–channel bidirectional chaos communication based on polarization–resolved chaos synchronization between twin 1550 nM VCSELs subject to variable–polarization optical injection. *Opt. Commun.* **2015**, *334*, 214–221. [CrossRef]
21. Klebanoff, A.; Hastings, A. Chaos in three species food chains. *J. Math. Biol.* **1994**, *32*, 427–451. [CrossRef]
22. Qu, Z. Chaos in the genesis and maintenance of cardiac arrhythmias. *Prog. Biophys. Mol. Biol.* **2011**, *105*, 247–257. [CrossRef]
23. Abrego, F.J.; Moiola, J.L. Lyapunov exponents analysis applied to a hyperchaotic prey–predator model. *IEEE Lat. Am. Trans.* **2013**, *11*, 230–235. [CrossRef]
24. Cramer, J.A.; Booksh, K.S. Chaos theory in chemistry and chemometrics: A review. *J. Chemom.* **2006**, *20*, 447–454. [CrossRef]
25. Vaidyanathan, S. Adaptive synchronization of chemical chaotic reactors. *Int. J. ChemTech Res.* **2015**, *8*, 612–621.
26. Smaoui, N.; Karouma, A.; Zribi, M. Adaptive synchronization of hyperchaotic Chen systems with applications to secure communications. *Int. J. Innov. Comput. Inf. Control* **2013**, *9*, 1127–1144.
27. Chandrasekaran, J.; Thiruvengadam, S.J. Ensemble of chaotic and naive approaches for performance enhancement in video encryption. *Sci. World J.* **2015**, *2015*, 458272. [CrossRef]
28. Naderi, B.; Kheiri, H. Exponential synchronization of chaotic system and application in secure communication. *Optik* **2016**, *127*, 2407–2412. [CrossRef]
29. Matsumoto, T. A chaotic attractor from Chua's circuit. *IEEE Trans. Circuits Syst.* **1984**, *31*, 1055–1058. [CrossRef]
30. Cuomo, K.; Oppenheim, A. Circuit Implementation of Synchronized Chaos with Application to Communication. *Phys. Rev. Lett.* **1993**, *71*, 65–68. [CrossRef]
31. Ma, Y.; Li, Y.; Jiang, X. Simulation and Circuit Implementation of 12–Scroll Chaotic System. *Chaos Solit. Fractals* **2015**, *75*, 127–133. [CrossRef]
32. Ranjan, R.; Mishra, S.; Madhekar, S. Electronic Circuit Implementation of Chaos Synchronization. *Eur. Phys. J. Spec. Top.* **2013**, *222*, 745–750. [CrossRef]
33. Fan, T.; Tuo, X.; Li, H.; He, P. Chaos Control and Circuit Implementation of Double–Wing Chaotic System. *Int. J. Numer. Model. Electron. Netw. Devices Fields* **2019**, *32*, e2611. [CrossRef]

34. Tlelo-Cuautle, E.; Rangel-Magdaleno, J.J.; Pano-Azucena, A.D.; Obseo-Rodelo, P.J.; Nunez-Perez, J.C. FPGA Realization of Multi–Scroll Chaotic Oscillators. *Commun. Nonlinear Sci. Numer. Simul.* **2015**, *27*, 66–80. [CrossRef]
35. Wang, Q.; Yu, S.; Li, C.; Fang, X.; Guyeux, C.; Bahi, J.M. Theoritical design and FPGA–based implementation of higher–dimensional digital chaotic systems. *IEEE Trans. Circuits Syst.* **2016**, *63*, 401–412. [CrossRef]
36. Tlelo-Cuautle, E.; Díaz-Muñoz, J.D.; González-Zapata, A.M.; Li, R.; León-Salas, W.D.; Fernández, F.V.; Guillén-Fernández, O.; Cruz-Vega, I. Chaotic Image Encryption Using Hopfield and Hindmarsh–Rose Neurons Implemented on FPGA. *Sensors* **2020**, *20*, 1326. [CrossRef] [PubMed]
37. Čelikovský, S.; Chen, G. On a generalized Lorenz canonical form of chaotic systems. *Int. J. Bifurc. Chaos* **2002**, *12*, 1789–1812. [CrossRef]
38. Campos-Cantón, E.; Barajas-Ramírez, J.G.; Solís-Perales, G.; Femat, R. Multiscroll attractors by switching systems. *Chaos Interdiscip. J. Nonlinear Sci.* **2010**, *20*, 013116. [CrossRef] [PubMed]
39. Campos-Cantón, E.; Femat, R.; Chen, G. Attractor Generated from Switching Unstable Dissipative Systems. *Chaos Interdiscip. J. Nonlinear Sci.* **2012**, *22*, 033121. [CrossRef]
40. Lorenz, E.N. Deterministic nonperiodic flow. *J. Atmos. Sci.* **1963**, *20*, 130–141. [CrossRef]
41. Rössler, O.E. An Equation for Continuous Chaos. *Phys. Lett. A* **1976**, *57*, 397–398. [CrossRef]
42. Chen, G.; Ueta, T. Yet Another Chaotic Attractor. *Int. J. Bifurc. Chaos* **1999**, *9*, 1465–1466. [CrossRef]
43. Díaz-González, E.C.; Aguirre-Hernández, B.; López-Rentería, J.A.; Campos-Cantón, E.; Loredo-Villalobos, C.A. Stability and multiscroll attractors of control systems via the abscissa. *Math. Probl. Eng.* **2017**, *2017*, 1–9. [CrossRef]

© 2020 by the authors. Licensee MDPI, Basel, Switzerland. This article is an open access article distributed under the terms and conditions of the Creative Commons Attribution (CC BY) license (http://creativecommons.org/licenses/by/4.0/).

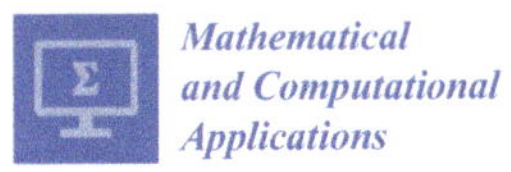

Article

Analysis and Comparison of Fuzzy Models and Observers for DC-DC Converters Applied to a Distillation Column Heating Actuator

Mario Heras-Cervantes [1], Adriana del Carmen Téllez-Anguiano [1,*], Juan Anzurez-Marín [2] and Elisa Espinosa-Juárez [2]

[1] Instituto Tecnológico de Morelia, Tecnológico Nacional de México, Av. Tecnológico 1500, Col. Lomas de Santiaguito, C.P. 58120 Morelia, Mich., Mexico; mario.hc@morelia.tecnm.mx

[2] Faculty of Electrical Engineering, Universidad Michoacana de San Nicolas de Hidalgo, Ciudad Universitaria, C.P. 58030 Morelia, Mich., Mexico; janzurez@umich.mx (J.A.-M.); eejuarez@umich.mx (E.E.-J.)

* Correspondence: adrianat@itmorelia.edu.mx

Received: 30 July 2020; Accepted: 26 August 2020; Published: 28 August 2020

Abstract: In this paper, as an introduction, the nonlinear model of a distillation column is presented in order to understand the fundamental paper that the column heating actuator has in the distillation process dynamics as well as in the quality and safety of the process. In order to facilitate the implementation control strategies to maintain the heating power regulated in the distillation process, it is necessary to represent adequately the heating power actuator behavior; therefore, three different models (switching, nonlinear and fuzzy Takagi–Sugeno) of a DC-DC Buck-Boost power converter, selected to regulate the electric power regarding the heating power, are presented and compared. Considering that the online measurements of the two main variables of the converter, the inductor current and the capacitor voltage, are not always available, two different fuzzy observers (with and without sliding modes) are developed to allow monitoring the physical variables in the converter. The observers response is compared to determine which has a better performance. The role of the observer in estimating the state variables with the purpose of using them in the sensors fault diagnosis, using the analytical redundancy concept, likewise, from the estimation of these variables other non-measurable can be determined; for example, the caloric power. The stability analysis and observers gains are obtained by linear matrix inequalities (LMIs). The observers are validated by MATLAB simulations to verify the observers convergence and analyze their response under system disturbances.

Keywords: distillation column heating actuator; Buck-Boost converter; Takagi–Sugeno model; fuzzy observer with sliding modes

---

1. Introduction

Nowadays, industrial processes have become increasingly complex to meet the quality and speed requirements that society demands. These requirements are reflected in a greater number of components and tasks that must have greater reliability, so they must comply with characteristics such as greater tolerance to failures and wear derived from their continuous operation, which implies the need to have control techniques that allow verifying and monitoring the even under fault conditions.

However, in the control of industrial processes, the online estimation of the variables that are not directly measurable is a fundamental problem, requiring measuring secondary variables that provide information to continuously monitor the status of the process.

One of the control elements used to carry out continuous monitoring of a system, which does not require investing in additional sensors, is the state observer. The observer operation is based on a mathematical model of the system and the information is obtained to reconstruct the variables that cannot be directly or easily measured. State observers can be used in several industrial applications.

One of the most widely used industrial processes today is distillation. Distillation is the process used to separate the components that form miscible liquid mixtures since 800 BC. Distillation is highly used in the chemical, petrochemical, food, pharmaceutical, and perfume industries. Currently, the need to produce chemical substances, such as ethanol, with an increasing demand to satisfy the need for disinfecting products, as well as the generation of biofuels that minimize damage to the environment, has incremented the need for optimal performance of the distillation process.

The distillation column is the most widely used equipment to carry out the distillation process. The operation of a distillation column involves working not only with chemicals but also under pressure and temperature conditions that may result in a risk for the user and the system if there is inadequate monitoring of the process. Figure 1 shows a simplified diagram of the distillation column.

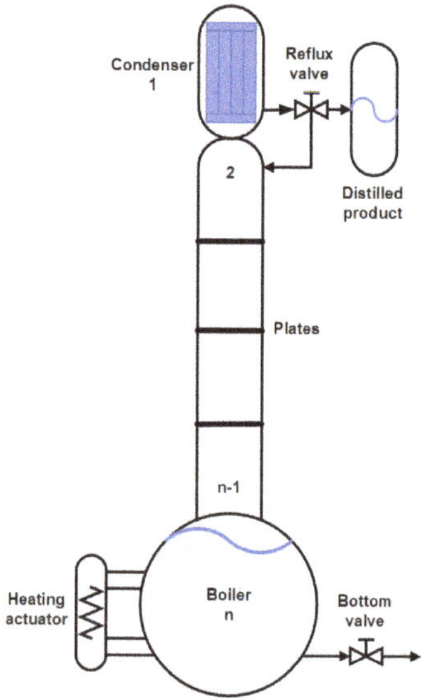

Figure 1. Distillation column simplified diagram.

Simulation is a key technology in the design, analysis, and operation of distillation columns. The reliability of simulations to represent the real process strongly depends on model quality, which needs to be reliable and predictive [1].

In the literature, modeling and control techniques, such as observers and control systems, have been applied to distillation columns to obtain a better analysis and understanding of the dynamics of the process, improving the quality of the distilled product and enhancing the user safety.

For instance, in [2], the authors model the phase equilibrium in biodiesel production by reactive distillation, The model is validated against experimental results for biodiesel production by reactive distillation. In [3], a surrogate-based optimization of distillation columns is investigated, the authors propose an implicit surrogate to cope with output multiplicities of the model.

In [4], the authors cope with modeling distillation columns with unknown parameters by proposing an intelligent, auto-regressive, exogenous Laguerre (AI-ARX-Laguerre) technique, which achieves average estimation accuracy improvements of 16% and 9% compared with the ARX and ARXLaguerre techniques, respectively.

In [5], a geometric observer (GO) based on the data from the Automatic Continuous Online Monitoring of Polymerization reactions (ACOMP) system for a semi-batch free radical polymerization reactor is presented.

In [6], a distributed high-gain observer design for a binary distillation process and another chemical process is presented and validated.

However, although there exist different works regarding the design and validation of distillation column models and observers, the importance of the related actuators is not always considered.

Actuators in a distillation column are fundamental in the process dynamics since they can modify physical process variables, such as temperature and pressure [7], modifying the purity of the product from reflux [8] or the speed of distillation.

The boiler actuator provides the amount of heat that is necessary to heat and evaporate the mixture to be distilled. This actuator generally controls the heat output by controlling electrical power. The amount of caloric power generated allows for controlling the speed for distillation in the process.

There are different ways to generate heat in a distillation column boiler, some of them use combustion appliances, such as natural gas, propane, and oil furnaces, which are cheap but difficult to control. An alternative is using heating resistors (Figure 2), which generated heat can be controlled by using an electronic power converter that can regulate the electric power that feeds the resistor.

Figure 2. Distillation column heating actuator.

As in the distillation columns, adequate modelling can facilitate the development of a control task, allowing to monitor the process dynamics by using state observers.

Different power converters models have been reported in the literature. Authors in [9] present a small-signal model for a single-stage PV fed Buck converter that acts as a battery charge-controller in order to improve its MPPT performance. In [10], a multi-physics model of Building-integrated photovoltaics (BIPV) integrated DC/DC converter is developed to quantify the potential of BIPV.

Other authors have reported models of different power converters such as a bidirectional DC/DC Buck [11], switching converters with power semiconductor filters [12] and a three-level T-type grid-connected converter system [13].

Additionally, different observers related to power converters have been reported. In [14], the authors estimate the power electronics modules by using a reduced-order state-space observer. A disturbance estimator to minimize the estimation error is coupled to the observer.

The developed observers can be used combined with other control structures, such as the observer developed in [15], where the authors estimate capacitor voltages from a modular multilevel converter and correct the estimation error by using a sliding mode control.

In [16], a Luenberger observer used as a residual generator is presented. The residuals help to detect open circuit faults in the power transistors of a DC-DC converter applied to a fuel cell. The inductor current is selected as a diagnostic variable to avoid the use of additional sensors to the system.

In [17], the design and simulation of a fault detection system based on a Luenberger observer are presented. This observer is used as a residual generator applied in DC-DC converter with Boost topology. The system detects faults in the voltage and current sensors used in the converter. The proposed system can reconfigure the converter output through the residual analysis.

In this paper, a Takagi–Sugeno fuzzy model that represents a Buck-Boost converter, used to regulate the heating power in a distillation column boiler is presented. The heating power is selected due to its impact in the thermal performance of the distillation process; when the thermal behavior of the columns is unstable can derive in several problems, for instance, a thermal shock to the boiler mixture can cause a violent siphon effect, affecting the measurements, the rate of the distilled product, and even the security of the process.

The developed Takagi–Sugeno fuzzy model is based on four fuzzy rules corresponding to two maximum and minimum operating points selected from the simulation of the case study. The Takagi–Sugeno model is simulated in Matlab and compared to the corresponding reduced nonlinear model of the converter in order to verify its performance.

Additionally, two different fuzzy observers based on the Takagi–Sugeno fuzzy model are presented and compared. The Takagi–Sugeno fuzzy observers difference is the inclusion or absence of a sliding mode term. The observer gains are obtained by linear matrix inequalities (LMIs), which are solved using mathematical software according to the methodology presented by [18].

2. Case Study: Distillation Column Heating Actuator

A distillation column is made up of three main parts: a condenser, a boiler, and the column body consisting of $n-2$ perforated plates. The vapor flow ascends through the plates of the column body enriching the light element (the element with the lowest boiling point of the mixture).

The vapor that reaches the condenser is condensed and, according to the state of the reflux valve, extracted as a distilled product or returned to the column. The returned liquid descends by gravity in the body of the column, enriching the heavy element (the element with the highest boiling point). Each plate in the distillation column corresponds to a degree of purity of the light element, known as the mole fraction.

Fractional distillation is used to separate homogeneous liquid mixtures in which the difference between the boiling points of the components is less than 25 °C. Each of the separated components is called a fraction.

In a distillation column, it is essential to continuously monitor all of the process variables to guarantee the quality and quantity of the distilled product, as well as the safety of the process and operators. To achieve this objective, it is necessary to have adequate control techniques.

2.1. Distillation Column Nonlinear Model

The mathematical model of a distillation column consists of a set of differential equations that represent the dynamics of each plate in the column in a stable state, i.e., when the first drop is distilled. Generally, the model of a distillation column is based on the balance of the light component in all of the plates, as shown in (1).

$$\frac{dx_i}{dt} = \frac{V(y_{i+1} - y_i) + L(x_{i-1} - x_i)}{M_i} \tag{1}$$

where V is the molar vapor flow, L the molar liquid flow, M_i the retained mass in plate i, x_i the liquid composition in plate i, y_i the vapor composition in plate i, x_{i+1} the liquid composition in plate $i-1$, y_{i+1} the vapor composition in plate $i-1$, x_{i-1} the liquid composition in plate $i-1$, and y_{i+1} the vapor composition in plate $i+1$, with $x, y \in \mathbb{R} : 0 < \mathrm{x} \leq 1, 0 < \mathrm{y} \leq 1$.

The diagram of the condenser is shown in Figure 3, its dynamics are expressed by the Equation (2), and the condenser is denominated as plate 1.

$$\frac{dx_1}{dt} = \frac{Vy_2 - Lx_1 - Dx_1}{M_1} \tag{2}$$

where M_1 is the retained mass in the condenser, x_1 the liquid composition in the condenser, y_1 the vapor composition in the condenser, x_2 the liquid composition in plate 2, y_2 the vapor composition in plate 2, and D the distilled product.

Figure 3. Distillation column condenser.

Figure 4 shows the scheme of a plate in the body column as well as the variables that are involved in the plate dynamics, expressed in Equation (3).

$$\frac{dx_i}{dt} = \frac{Vy_{i+1} - Vy_i + Lx_{i-1} - Lx_i}{Mi} \tag{3}$$

with $i = 2, 3, \ldots\ n - 1$.

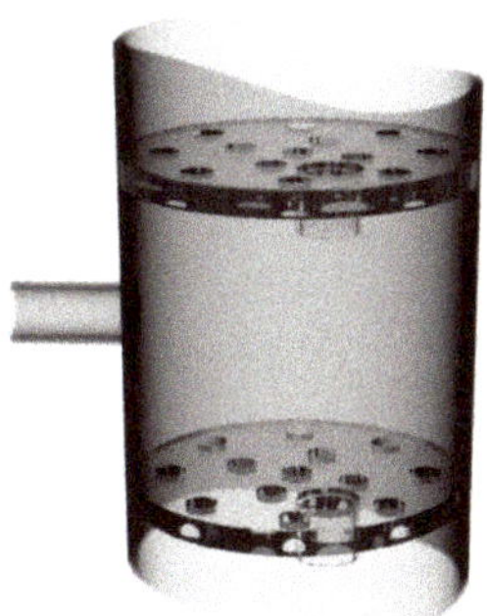

Figure 4. Plates in the column body.

Figure 5 shows a scheme of the boiler in a distillation column, denominated plate n, its dynamics is presented in (4).

$$\frac{dx_n}{dt} = \frac{Vx_n - Vy_n + Lx_{n-1} - Lx_n}{M_n} \tag{4}$$

where M_n is the retained mass in the boiler, x_n the liquid composition in the boiler, y_n the vapor composition in the boiler, x_{n-1} the liquid composition in plate $n - 1$, and n the total number of plates.

Figure 5. Distillation column boiler.

Besides, the molar flows are considered in the model, in a batch distillation column these flows are [19]: vapor molar flow (5), liquid molar flow (6), and distilled product (7).

$$V = \frac{Qb}{H_i^{vap} x_n + H_j^{vap}(1 - x_n)} \tag{5}$$

$$L = (1 - Rf)V \tag{6}$$

$$D = V - L \tag{7}$$

where Qb is the heating power, Rf the reflux, H_i^{vap} the vapor enthalpy of the light component in the mixture, and H_j^{vap} the vapor enthalpy of the heavy component in the mixture.

As can be seen, the heating power determines the distillation column dynamics, thus it is essential to have an adequate representation of its behavior.

2.2. Heating Actuator (Boiler) Model

The heating power is determined by the Joule law of heating [20] expressed as:

The power of heating generated by an electrical conductor is proportional to the product of its resistance and the square of the electric current passing through the conductor and the time the current flows through the wire.

The Joule law is defined by (8)

$$J = i^2 Rt \tag{8}$$

where J is the generated heat in joules, i the electrical current in amperes, R the resistance in ohms and t the time in seconds.

The law of conservation of energy affirms that energy cannot be created or destroyed, it can only be changed from one form to another. Joule's Law expressed in Electric Power (P) is expressed in (9):

$$w = Pt = Qb \tag{9}$$

The heating resistance converts electrical energy into heat by circulating current in the conductor; therefore, the thermal power of a boiler can be manipulated and modeled on the electrical power dissipated by the resistance.

Figure 6 shows the actuator scheme, which adjusts the power in the boiler heating resistor by regulating the voltage with a DC-DC converter. This is a simplified model of the buck-boost converter [21], widely used for the analysis of its operation and control [22–26]. DC-DC converters can regulate the output voltage to the desired value from the switching of electronic devices, usually diodes and transistors. Differences with the actual behavior of the converter are compensated by the further inclusion of a state observer.

These power electronics converters have applications in renewable energy systems, smart grids, and home and laboratory equipment power systems [27,28]. The basic topologies of DC-DC converters are Buck, Boost, and Buck-Boost [21]. The Buck converter is characterized by the output voltage being lower than the input voltage, the Boost converter is characterized by the output voltage being greater than the input voltage and the Buck-Boost converter is a step-down converter, depending on the cycle (d). For d less than 0.5, the converter reduces the voltage and for d more than 0.5 the voltage is increased.

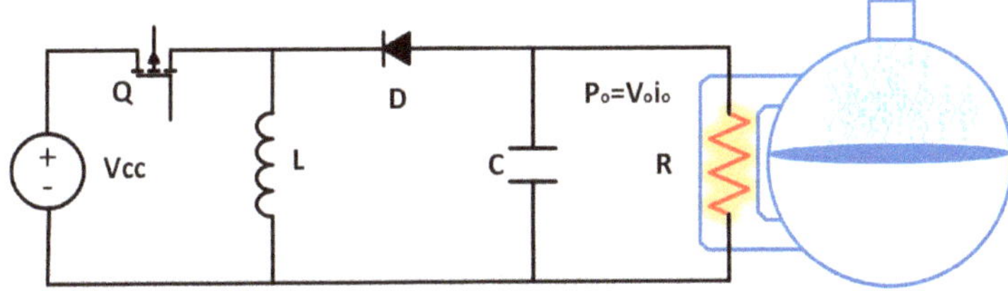

Figure 6. Scheme of the heating power actuator.

The operating principle of DC-DC converters is the switching of topological states.

3. Buck-Boost Converter Models

3.1. State-Space Model of the Buck-Boost Converter

The model of the Buck-Boost converter is obtained from commuting the Q switch on and off, which is, from the set of expressions obtained by each topological state. The system is represented as the states matrix form (10) taking as state variables the inductor current (i_L) and the capacitor voltage (v_C).

$$\dot{x}(t) = Ax(t) + Bu(t) \tag{10}$$

During the on time ($t = ON$), the converter is described by (11):

$$\dot{x}(t) = \begin{pmatrix} 0 & 0 \\ 0 & -\frac{1}{RC} \end{pmatrix} \begin{pmatrix} i_L \\ v_C \end{pmatrix} + \begin{pmatrix} \frac{1}{L} \\ 0 \end{pmatrix} V_{cc} \tag{11}$$

During the off time ($t = OFF$), the converter is described by (12):

$$\dot{x}(t) = \begin{pmatrix} 0 & \frac{1}{L} \\ -\frac{1}{C} & -\frac{1}{RC} \end{pmatrix} \begin{pmatrix} i_L \\ v_C \end{pmatrix} + \begin{pmatrix} 0 \\ 0 \end{pmatrix} V_{cc} \tag{12}$$

The resulting model of the Buck-Boost converter is a commuting model between two linear subsystems, represented in (11) and (12), which commutation depends on the switch state Q. The general representation of the system in matrix form (10) is expressed in (13).

$$\dot{x}(t) = A_k x(t) + B_k u(t) \tag{13}$$

where k is the subsystem for each switch state: $k = 1, 2$

$$A_1 = \begin{pmatrix} 0 & 0 \\ 0 & -\frac{1}{RC} \end{pmatrix}$$
$$A_2 = \begin{pmatrix} 0 & \frac{1}{L} \\ -\frac{1}{C} & -\frac{1}{RC} \end{pmatrix}$$

$$B_1 = \begin{pmatrix} \frac{1}{L} \\ 0 \end{pmatrix}$$

$$B_2 = \begin{pmatrix} 0 \\ 0 \end{pmatrix}$$

$$x = \begin{pmatrix} i_L \\ v_C \end{pmatrix}$$

Switching Model Simulation

The parameters of the electrical variables of converter operation, considered as a specific case study, are presented in Table 1.

Table 1. Buck-Boost converter operation parameters.

Parameter	Magnitude
Input voltage V_{in}	180 V
Output voltage v_C	−229 V
Inductor current i_L	7.3 A
Switching frequency f	20 kHz
Load R	70.3 Ω
Inductor L	5 mH
Capacitor C	78 µF
Duty cycle d	0.56

Figures 7 and 8 show the simulation of the Buck-Boost converter dynamics, as well as the output ripple, when considering initial conditions in the state variables equal to zero ($\dot{x}(0) = 0$).

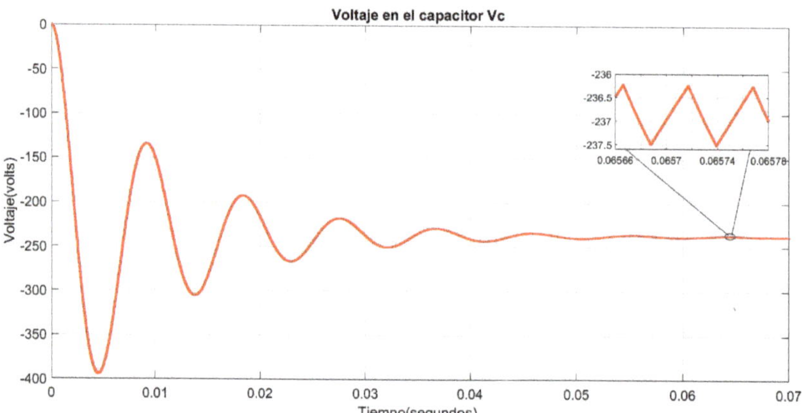

Figure 7. Capacitor voltage in the Buck-Boost switching model.

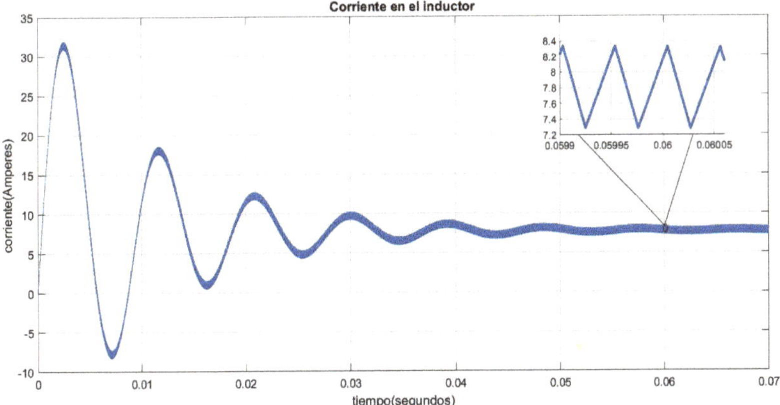

Figure 8. Inductor current in the Buck-Boost switching model.

3.2. Nonlinear Model

The nonlinear model of the converter involves unifying the linear subsystems, including the control variable u, which is determined by the duty cycle and takes values between 0 and 1, as shown in (14):

$$\dot{x}(t) = (A_1 x + b_1)d + (A_2 x + b_2)(1 - d) \tag{14}$$

or equivalently in (15):

$$\dot{x}(t) = A_2 x + b_2 + (A_1 - A_2)xd + (b_1 - b_2)d \tag{15}$$

The model response is an average of both linear submodels (11) and (12).

$$\dot{x}(t) = \begin{pmatrix} 0 & \frac{1}{L} \\ -\frac{1}{C} & -\frac{1}{RC} \end{pmatrix} \begin{pmatrix} x_1 \\ x_2 \end{pmatrix} + \begin{pmatrix} \frac{V_{cc} - x_2}{L} \\ \frac{x_1}{C} \end{pmatrix} d \tag{16}$$

3.3. Takagi–Sugeno Fuzzy Model

The Takagi–Sugeno Fuzzy Model has an important advantage, because it analyzes the operation range of nonlinear systems, using the local sector concept; this means that it takes advantage of the characteristic of the physical systems which are bounded, which allows for modeling the nonlinear systems through a linear subsystems set.

Based on the nonlinear model of the Buck-Boost converter presented in (16) and considering as fuzzy variables the states ($z_1 = v_C, z_2 = i_L$) that operate between maximum and minimum nominal values ($v_{C_{max}}, v_{C_{min}}, i_{L_{max}}, i_{L_{min}}$), a Takagi–Sugeno (T-S) fuzzy model that interpolates between four linear submodels based on the following rules is proposed:

Rule 1:
If z_1 is $z_{1_{min}}$ and if z_2 is $z_{2_{min}}$
Then:
$\dot{x}(t) = A_1 x(t) + B_1 d$

Rule 2:
If z_1 is $z_{1_{min}}$ and if z_2 is $z_{2_{max}}$
Then:
$\dot{x}(t) = A_2 x(t) + B_2 d$

Rule 3:
If z_1 is $z_{1_{max}}$ and if z_2 is $z_{2_{min}}$
Then:
$$\dot{x}(t) = A_3x(t) + B_3d$$

Rule 4:
If z_1 is $z_{1_{max}}$ and if z_2 is $z_{2_{max}}$
Then:
$$\dot{x}(t) = A_4x(t) + B_4d$$

According to the converter characteristics, the linear submodels are obtained while using the nonlinear sector ($z = [max, min]$). Where:

$$A_1 = \begin{pmatrix} 0 & \frac{1}{L} \\ -\frac{1}{C} & -\frac{1}{RC} \end{pmatrix} = A_2 = A_3 = A_4 = A$$

$$B_1 = \begin{pmatrix} \frac{V_{in}+z_{1_{min}}}{L} \\ \frac{z_{2_{min}}}{C} \end{pmatrix}, B_2 = \begin{pmatrix} \frac{V_{in}+z_{1_{min}}}{L} \\ \frac{z_{2_{max}}}{C} \end{pmatrix}, B_3 = \begin{pmatrix} \frac{V_{in}+z_{1_{max}}}{L} \\ \frac{z_{2_{min}}}{C} \end{pmatrix}, B_4 = \begin{pmatrix} \frac{V_{in}+z_{1_{max}}}{L} \\ \frac{z_{2_{max}}}{C} \end{pmatrix}$$

$$C_1 = \begin{pmatrix} 1 & 0 \\ 0 & 1 \end{pmatrix} = C_2 = C_3 = C_4 = C$$

The membership functions ($\mu(z)$) for the fuzzy sets are determined by Equation (17) for the capacitor voltage $v_C = z_1$:

$$\mu(z_1) = \begin{cases} \mu_{z_{1_{min}}} & = \frac{z_{1_{max}} - z_1}{z_{1_{max}} - z_{1_{min}}} \\ \mu_{1_{max}} & = 1 - \mu_{z_{1_{min}}} \end{cases} \tag{17}$$

and Equation (18) for the inductor current $i_L = z_2$:

$$\mu(z_2) = \begin{cases} \mu_{z_{2_{min}}} & = \frac{z_{2_{max}} - z_2}{z_{2_{max}} - z_{2_{min}}} \\ \mu_{z_{2_{max}}} & = 1 - \mu_{z_{2_{mins}}} \end{cases} \tag{18}$$

The normalized weights are given by (19):

$$\begin{aligned} h_1(z_1, z_2) &= \mu_{z_{1min}}\mu_{z_{2min}} \\ h_2(z_1, z_2) &= \mu_{z_{1min}}\mu_{z_{2max}} \\ h_3(z_1, z_2) &= \mu_{z_{1max}}\mu_{z_{2min}} \\ h_4(z_1, z_2) &= \mu_{z_{1max}}\mu_{z_{2max}} \end{aligned} \tag{19}$$

The T-S fuzzy model for the DC-DC converter is given in (20) when considering $r = 4$.

$$\begin{aligned} \dot{x}(t) &= Ax(t) + (\textstyle\sum_{i=1}^{r} h_i(z_1, z_2)B_i)d \\ y(t) &= \textstyle\sum_{i=1}^{r} h_i(z_1, z_2)C_ix(t) \end{aligned} \tag{20}$$

3.4. Models Comparison

In this section, the converter nonlinear and the Takagi–Sugeni models are compared to the response of the switching model in simulation.

In the models simulation in 0.1 s, the load is decreased to 78.94.88% ($R = 55.5\ \Omega$) of its nominal value, in 0.2 s it is increased to 113.79% ($R = 80\ \Omega$). In 0.3 s, two simultaneous disturbances are simulated, the load increases to 113.79% ($R = 80\ \Omega$) of its nominal value, and the input voltage is increased to 116.66 % ($V_{CC} = 210\ V$) of its nominal value.

Figure 9 shows the behavior of the capacitor voltage (v_C) as compared to the nonlinear and the Takagi–Sugeno models response.

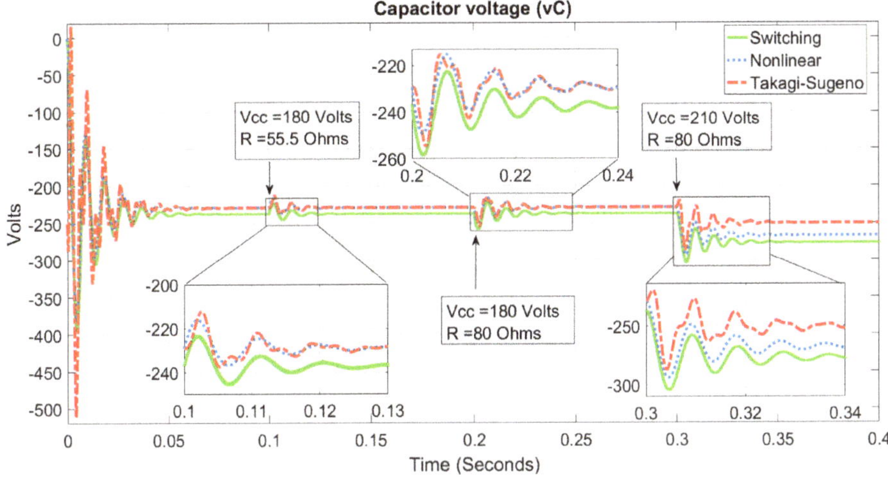

Figure 9. Comparison of the models capacitor voltage dynamics.

Figure 10 shows the behavior of the inductor current (i_L) as compared to the nonlinear and the Takagi–Sugeno models response.

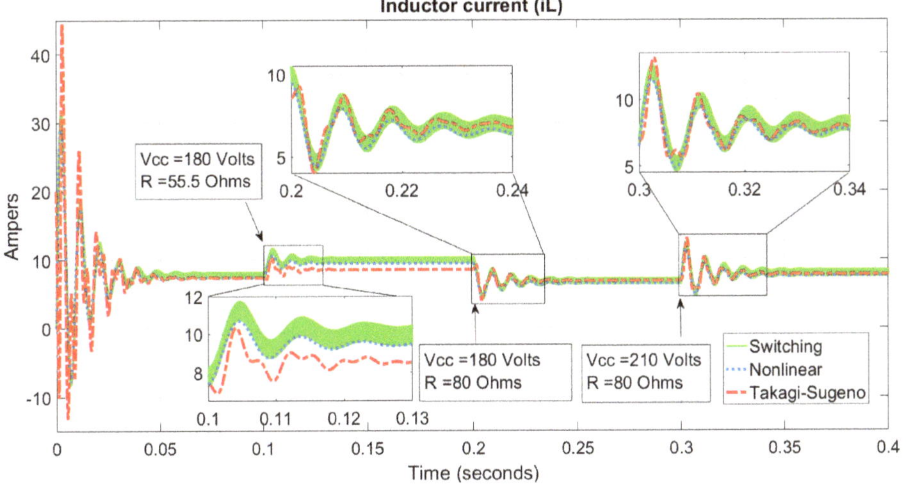

Figure 10. Comparison of the models inductor current dynamics.

The Takagi–Sugeno model has an adequate response as long as the system dynamics are constrained into the operating points used for its design.

4. Observers

A state observer is a dynamic system that estimates state variables or parameters from available measurements. Observers, also called virtual sensors, are widely used because they allow estimating difficult-to-measure variables of a system from available mathematical algorithms and measurements, and they are also suitable for detecting and locating faults in actuators and sensors.

Observers have a systematic and simple design procedure, facilitating their implementation and execution in real time. The mathematical model is a fundamental part of an observer since it allows describing the dynamics of a real system.

Given a system expressed in the form (10), the observer general equation for reconstructing or estimating states of a system is described in (21):

$$\dot{\hat{x}}(t) = \underbrace{A\hat{x} + Bu(t)}_{Predictor} + \underbrace{L(y(t) - \hat{y}(t))}_{Corrector} \tag{21}$$

where $\hat{x} \in \Re$ represents the estimated state during time $\tau > t_0$ and the estimated output is defined in (22):

$$\hat{y}(t) = C\hat{x}(t) \tag{22}$$

The system that is presented in (21) and (22) is also denoted as observer Luenberger identity and it is coupled to the original process through the inputs and outputs, as shown in Figure 11. The observer consists of two parts: a predictive stage, based on the model of the observed system, and a corrective stage, formed by the estimation error defined by the difference between the real output $y(t)$ and the estimated output $\hat{y}$, expressed in (23):

$$e(t) = y(t) - \hat{y}(t) \tag{23}$$

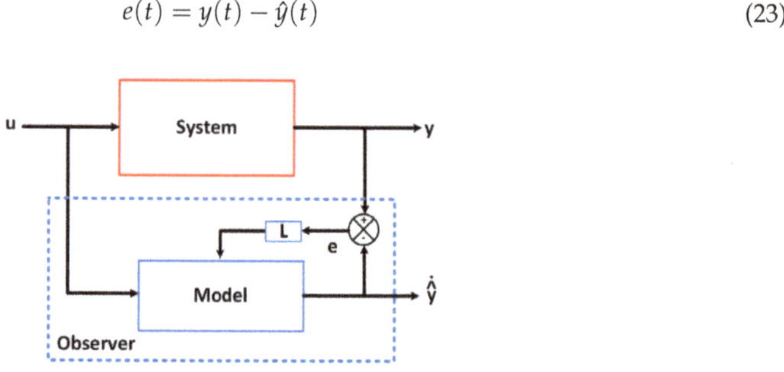

Figure 11. State observer general scheme.

4.1. Fuzzy Observer

Combining the Takagi–Sugeno fuzzy model of a nonlinear system expressed in (24) with the Luenberger observer (21), the general structure of a fuzzy observer is obtained [29].

$$\begin{gathered} \dot{\hat{x}}(t) = \sum_{i=1}^{r} h_i(z(t))[A_i\hat{x}(t) + B_iu(t) + K_i\tilde{e}] \\ \hat{y}(t) = \sum_{i=1}^{r} h_i(z(t))C_i\hat{x}(t) \end{gathered} \tag{24}$$

Tanaka, in [18], demonstrates the stability of the fuzzy observer as long as there is a P matrix that satisfies the LMI's expressed in (25):

$$\begin{gathered} P > 0 \\ N_i > 0 \\ A_i^T P - C_i^T N_i^T + PA_i - N_iC_i < 0 \\ A_i^T P - C_j^T N_i^T + PA_i - N_iC_j + PA_j^T - C_i^T N_j^T + PA_j - N_jC_i < 0 \\ i < j \end{gathered} \tag{25}$$

4.2. Robust Observer under Parameter Variation

The general structure of the observer with sliding modes proposed in [30], is expressed in (26)

$$\begin{aligned} \dot{\hat{x}}(t) &= A\hat{x}(t) + Bu(t) + K(y(t) - \hat{y}(t)) + \varphi(t) \\ \hat{y}(t) &= C\hat{x}(t) \end{aligned} \tag{26}$$

where, $\varphi(t)$ is the sliding-mode discontinuous vector.

For the convergence analysis, the estimation error $e(t)$ is defined in (27).

$$e(t) = x(t) - \hat{x}(t) \tag{27}$$

The error dynamics $\dot{e}(t)$ is defined in (28)

$$\begin{aligned} \dot{e}(t) &= \dot{x}(t) - \dot{\hat{x}}(t) \\ &= Ax(t) + Bu(t) + Ed(t) - A\hat{x}(t) - Bu(t) - K(y(t) - \hat{y}(t)) - \varphi(t) \\ &= A(x(t) - \hat{x}(t)) + Ed(t) - KC(x(t) - \hat{x}(t)) - \varphi(t) \\ \dot{e}(t) &= \overline{A}e(t) + Ed(t) - \varphi(t) \end{aligned} \tag{28}$$

where $\overline{A}$ is defined in (29).

$$\overline{A} = (A - KC) \tag{29}$$

where the Lyapunov function $v(t)$ expressed in (30)

$$v(t) = e^T(t)P_{sm}e(t) \tag{30}$$

is derived (30) omitting the time dependence.

$$\begin{aligned} \dot{v} &= e^T P_{sm}\dot{e} + \dot{e}^T P_{sm}e \\ \dot{v} &= e^T P_{sm}[\overline{A}e + Ed - \varphi] + [\overline{A}e + Ed - \varphi]^T P_{sm}e \\ \dot{v} &= e^T P_{sm}\overline{A}e + e^T P_{sm}Ed - e^T P_{sm}\varphi + e^T\overline{A}P_{sm}e^T + d^T E^T P_{sm}e - \varphi^T P_{sm}e \\ \dot{v} &= e^T[P_{sm}\overline{A} + \overline{A}^T P_{sm}]e + 2e^T P_{sm}Ed - 2e^T P_{sm}\varphi \end{aligned} \tag{31}$$

and γ is defined as (32).

$$\gamma = P_{sm}\overline{A} + \overline{A}^T P_{sm} \tag{32}$$

Rewriting (31)

$$\dot{v} = \gamma||e||^2 + 2||e^T P_{sm}||||Ed|| - 2e^T P_{sm}\varphi \tag{33}$$

Because γ corresponds to a Lyapunov inequality, it is defined negative, so the $\gamma||e||^2$ is negative. To achieve the condition of $\dot{v} < 0$, it is proposed that φ has the following form.

$$\varphi^T = M\frac{e^T P_{sm}}{||e^T P_{sm}||} = Msign(e^T P_{sm}) \tag{34}$$

where M is a constant positive gain. P_{sm} is a defined positive matrix, which must fulfill the Lyapunov equation.

$$P_{sm}\overline{A} + \overline{A}^T P_{sm} < 0 \tag{35}$$

Omitting $\gamma||e||^2$ in Equation (33), since it always has a negative value.

$$\begin{aligned} 2||e^T P_{sm}||||Ed|| - 2e^T P_{sm}[M\frac{e^T P_{sm}}{||e^T P_{sm}||}] < 0 \\ 2||e^T P_{sm}||[||Ed|| - M] < 0 \\ ||Ed|| < M \end{aligned} \tag{36}$$

M value is selected large enough to satisfy the condition (36).

4.3. Fuzzy Observer with Sliding Modes

The fuzzy observer with sliding modes [31] is based on the Luenberger observer [32] for linear systems and the Tanaka fuzzy observer [29].

Using the fuzzy observer (24), it is possible to build local observers with sliding modes for each linear subsystem [30]. Each observer is associated with a fuzzy rule i, given as:

If:
$z_1(t)$ is $M1$ and, $\cdot$, and $z_p(t)$ is Mp ,

$$\textbf{Then}: \begin{cases} \dot{\hat{x}}(t) = A_i\hat{x}(t) + B_i u(t) + K_i(y(t) - \hat{y}(t)) + \varphi_i(t) \\ \hat{y}(t) = C_i\hat{x}(t) \end{cases}$$

The complete observer is given by the weighted sum of each subsystem.

$$\dot{\hat{x}}(t) = \sum_{i=1}^{N} h_i(z(t))[A_i\hat{x}(t) + B_i u(t) + K_i C_i(x(t) - \hat{x}(t)) + \varphi_i(t)] \tag{37}$$

$$\hat{y}(t) = \sum_{i=1}^{N} h_i(z(t))C_i\hat{x}(t) \tag{38}$$

$\varphi_i(t)$ is the discontinuous vector of sliding modes for the subsystem i, which has the following form:

$$\varphi(t)_i = M_i sign(P_i e(t)) \tag{39}$$

where $M_i > 0$ is a positive constant, $P_i > 0$ so the Lyapunov equation is fulfilled, and $e(t)$ is defined as the estimated state error.

$$e(t) = (x(t) - \widehat{x(t)}) \tag{40}$$

The stability of the complete observer is demonstrated if each pair A_i, C_i is observable and P_i fulfills the Lyapunov equation.

$$P_i\overline{A_i} + \overline{A_i}^T P_i < 0 \tag{41}$$

where,

$$\overline{A_i} = A_i - K_i C_i \tag{42}$$

The structure of the fuzzy observer with sliding modes for Takagi–Sugeno systems defined in [31,33] is expressed in (43).

$$\begin{aligned} \dot{\hat{x}}(t) = \textstyle\sum_{i=1}^{r} h_i(z(t))[A_i\hat{x}(t) + B_i u(t) + K_i\tilde{e} + \varphi_i(t)] \\ \hat{y}(t) = \textstyle\sum_{i=1}^{r} h_i(z(t))C_i\hat{x}(t) \end{aligned} \tag{43}$$

The fuzzy observer estimation error with sliding modes is determined by

$$\tilde{e}(t) = y(t) - \hat{y}(t) \tag{44}$$

and $\varphi_i(t)$ is the discontinuous vector of sliding modes for the subsystem i, when considering the estimation error as the sliding surface expressed in (45):

$$\varphi(t)_i = sign(\tilde{e}' P) \tag{45}$$

The stability of the fuzzy system expressed in (43) is obtained using quadratic Lyapunov functions of the form (46);

$$V(x(t)) = x^T P x(t) \tag{46}$$

Asymptotic stability is guaranteed if there is a defined positive P matrix ($P > 0$) such that fulfills the Lyapunov equation expressed in (47).

$$A_i^T P + PA_i < 0 \tag{47}$$

for each subsystem with $i = 1, 2, \cdots, r$.

The resulting linear matrix inequalities (LMI) that guarantee the stability of the fuzzy observer in all subsystems and their combinations are expressed in (48):

$$\begin{gathered} P > 0 \\ N_i > 0 \\ A_i^T P - C_i^T N_i^T + PA_i - N_i C_i < 0 \\ A_i^T P - C_j^T N_i^T + PA_i - N_i C_j + PA_j^T - C_i^T N_j^T + PA_j - N_j C_i < 0 \\ i < j \end{gathered} \tag{48}$$

Observer gains are defined by the LMI's system solution defined in (49)

$$K_i = P_i^{-1} N_i \tag{49}$$

5. Fuzzy Observer for the Heating Resistor Actuator

According to the fuzzy Takagi–Sugeno model for the boiler Buck-Boost converter defined in (20), the expression of the fuzzy observer with sliding modes for Buck-Boost converter is defined in (50).

$$\begin{gathered} \dot{\hat{x}} = A\hat{x}(t) + (\textstyle\sum_{i=1}^{4} h_i(z_1, z_2) B_i) d \\ \hat{y}(t) = \textstyle\sum_{i=1}^{4} h_i(z_1, z_2) C_i \hat{x}(t) \end{gathered} \tag{50}$$

The output matrix C is defined in (51), where the system outputs are v_C and i_L.

$$C = \begin{pmatrix} 1 & 0 \\ 0 & 1 \end{pmatrix} \tag{51}$$

The observer estimation error is defined by the difference of the nonlinear model states of the converter ($\dot{x}$) and the estimated states of the fuzzy observer ($\dot{\hat{x}}$) expressed in (52).

$$\tilde{e}_{\phi_a} = \dot{x} - \dot{\hat{x}} \tag{52}$$

The block diagram of the observer applied to the Buck-Boost converter is shown in Figure 12, where the fuzzy variables (z_1, z_2) are the states of the system (x_1, x_2), the gains of the fuzzy observer are defined as K_f and K_ϕ.

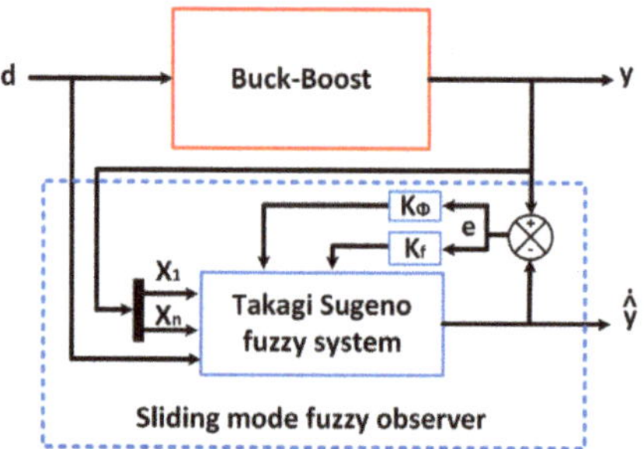

Figure 12. DC-DC converter fuzzy observer.

According to the characteristics of the converter fuzzy model of the converter (20), where the state matrices $A_1, A_2, A_3, A_4 = A$ are identical, the LMI's system to guarantee the stability of the fuzzy observer for the Buck-Boost converter is defined in (53).

$$\begin{array}{c} P_{\phi_a}\phi > 0 \\ A' P_{\phi_a} - C' N'_{\phi_a} + P_{\phi_a} A - N_{\phi_a} C < 0 \end{array} \tag{53}$$

Given the solution for P_{ϕ_a}, the gain K_ϕ for the observer is determined by (54)

$$K_\phi = P_{\phi_a}^{-1} N_{\phi_a} \tag{54}$$

6. Discussion, Analysis and Results

The observer simulation is performed for a Buck-Boost converter that regulates the voltage to a 350 W heating resistor for a distillation column boiler.

The observer design parameters are determined for the case study presented in Table 2; and, the behavior of observer is validated by simulation in MATLAB when considering the following disturbances: variation in input voltage to the converter caused mainly by the supply voltage (line voltage) and variations in load are usually due to degradation or manufacturing of the heating resistance.

According to the characteristics of the fuzzy system, where $A_1, A_2, A_3, A_4 = A$, the LMI's system used to determine the stability of the fuzzy system with eight closed-loop rules with observer is expressed in (55).

$$\begin{array}{c} P_1 > 0 \\ P_2 > 0 \\ A^T P_2 - C_1^T N_1^T + P_2 A - N_1 C_1 < 0 \\ P_1 A^T - M_1^T B_1^T + A P_1 - B_1 M_1 < 0 \\ P_1 A^T - M_2^T B_2^T + A P_1 - B_2 M_2 < 0 \\ P_1 A^T - M_3^T B_3^T + A P_1 - B_3 M_3 < 0 \\ P_1 A^T - M_4^T B_4^T + A P_1 - B_4 M_4 < 0 \end{array} \tag{55}$$

The LMI's that represent the overlaps of the membership functions are expressed in (56)

$$\begin{array}{c} P_1 A^T - M_2^T B_1^T + A P_1 - B_1 M_2 - M_1^T B_2^T - B_2 M_1 < 0 \\ P_1 A^T - M_3^T B_1^T + A P_1 - B_1 M_3 - M_1^T B_3^T - B_3 M_1 < 0 \\ P_1 A^T - M_4^T B_1^T + A P_1 - B_1 M_4 - M_1^T B_4^T - B_4 M_1 < 0 \\ P_1 A^T - M_3^T B_2^T + A P_1 - B_2 M_3 - M_2^T B_3^T - B_3 M_2 < 0 \\ P_1 A^T - M_4^T B_2^T + A P_1 - B_2 M_4 - M_2^T B_4^T - B_4 M_2 < 0 \\ P_1 A^T - M_4^T B_3^T + A P_1 - B_3 M_4 - M_3^T B_4^T - B_4 M_3 < 0 \end{array} \tag{56}$$

Given the solution for $P2$, the gain K for the observer is determined by (57)

$$K = P_2^{-1} N_1 \tag{57}$$

6.1. Simulation and Comparison of the Heating Actuator Observers

The observer simulation is performed for a Buck-Boost converter with the characteristics presented in Table 2.

Table 2. Parameters of the Buck-Boost converter.

Parameter	Magnitude
Supply voltage (V_{cc})	180 V
Output voltage (V_{out})	−229 V
Inductor (L)	5 µH
Capacitor (C)	78 µF
Load (R)	70.3 Ω
Switching frequency (f)	20 kHz
Duty cycle (u)	0.56

Figure 13 shows the simulation in the voltage transient in the capacitor (v_C) with different initial conditions of the observer and the nonlinear model ($v_C(0) = 0$ V, $\hat{v}_C(0) = -250$ V), where the convergence of the fuzzy observer ($\hat{x}_2$) is shown at 400 µs, and the convergence of the fuzzy observer with sliding modes is shown at 9 ms.

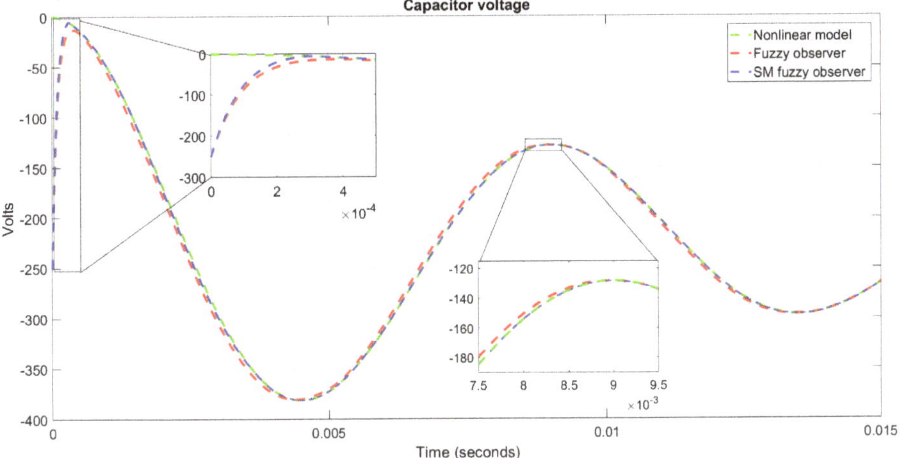

Figure 13. Voltage transient in v_C with different conditions, $v_C(0) = 0$ V and $\hat{v}_C(0) = -250$ V.

Figure 14 shows the transient in the inductor current (i_L) with different initial conditions between the observer and the nonlinear model ($i_L(0) = 0$ A and $\hat{i}_L(0) = 20$ A), where the convergence of the fuzzy observer is shown at 100 µs and the convergence of the fuzzy observer with sliding modes is shown at 14 ms.

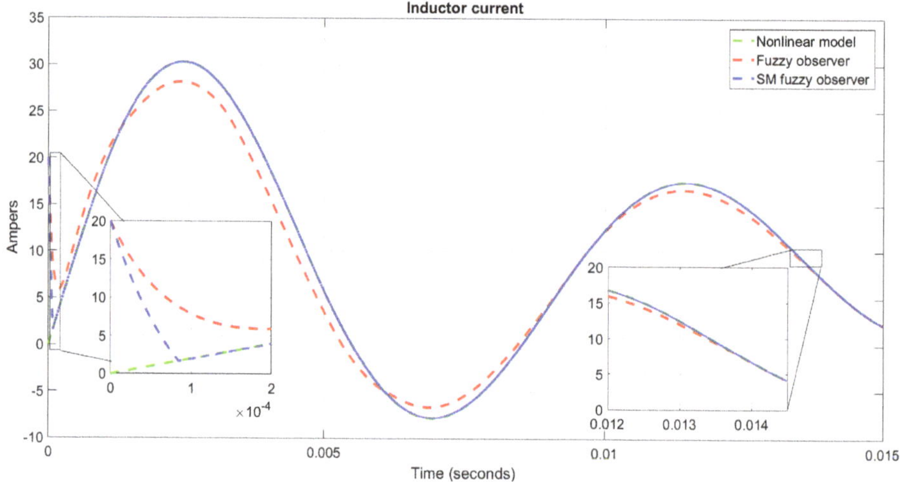

Figure 14. Current transient in v_C with different conditions, ($\hat{i}_L(0) = 0$ A, $\hat{v}_C(0) = 20$ A).

Figure 15 shows the convergence of the observer in the capacitor voltage v_C with disturbances in the nominal input voltage ($V_in = 180$ V). In the observer simulation in 0.1 s, the voltage V_{in} is decreased to 88.88% ($V_{in} = 160$ V) of its nominal value, in 0.2 s it is increased to 111.11% ($V_{in} = 200$ V), in both cases, the observer ($\hat{v}_C$) converges to the capacitor voltage (v_C) of the nonlinear model under these disturbances. The observer has a maximum estimation error of 1.3 V and a minimum error of 100 µV due to the chattering effect of the sliding surface.

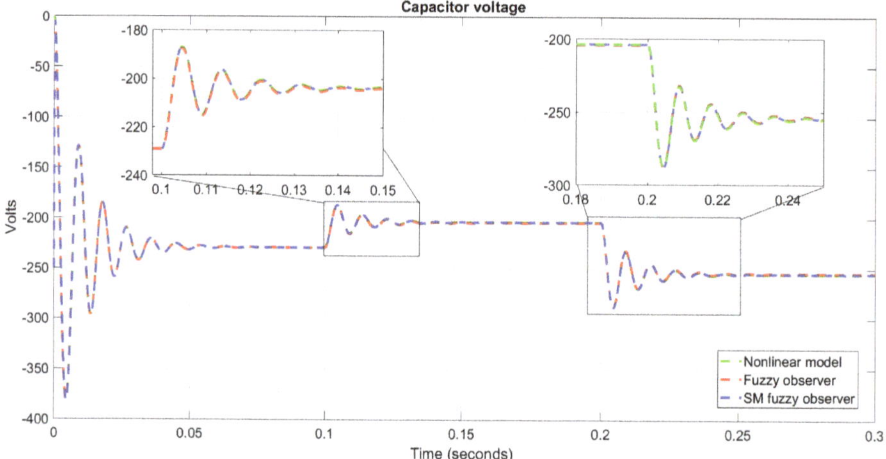

Figure 15. Observer response for the v_C voltage with disturbances in the supply voltage V_{in}.

Figure 16 shows the convergence of the observer in the inductor current i_L before disturbances in the nominal input voltage ($V_{in} = 180$ V). In 0.1 s the voltage V_{in} is decreased to 88.88% ($V_{in} = 160$ V) and, in 0.2 s, it is increased to 111.11% $V_{in} = 200$ V, in both cases the observer ($\hat{i}_L$) converges to the current in the inductor (i_L) of the nonlinear model before these disturbances. The observer presents a maximum estimation error of 394 mA and a minimum error of 80 µA.

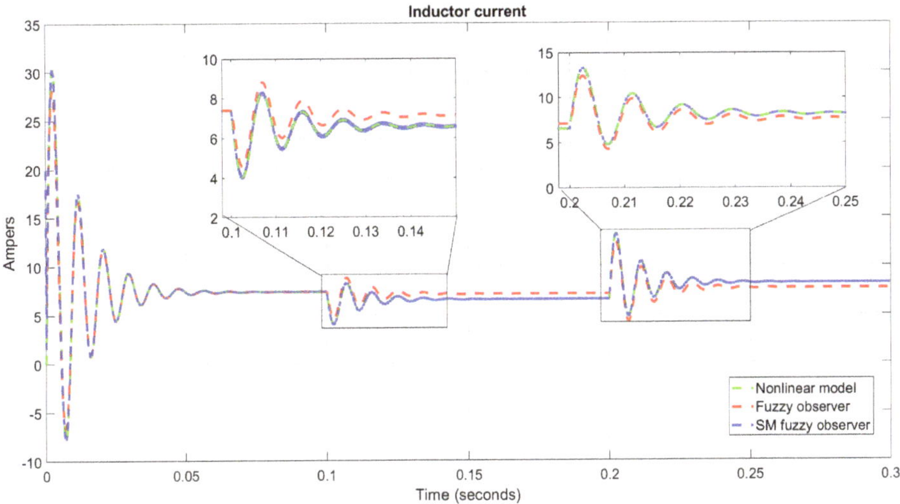

Figure 16. Observer response for the inductor current i_L with disturbances in supply voltage V_{in}.

Figure 17 shows the observer convergence in the capacitor voltage under variations in the nominal load ($R = 70.3\ \Omega$). In 0.1 s the load is decreased to 78.23% ($R = 55\ \Omega$) and in 0.2 s it is increased to 113.79% ($R = 8\ \Omega$), in both cases, the observer ($\hat{i}_L$) converges to the capacitor voltage (v_C) of the nonlinear model. The observer has a maximum error of 1 V and a minimum of 75 μV.

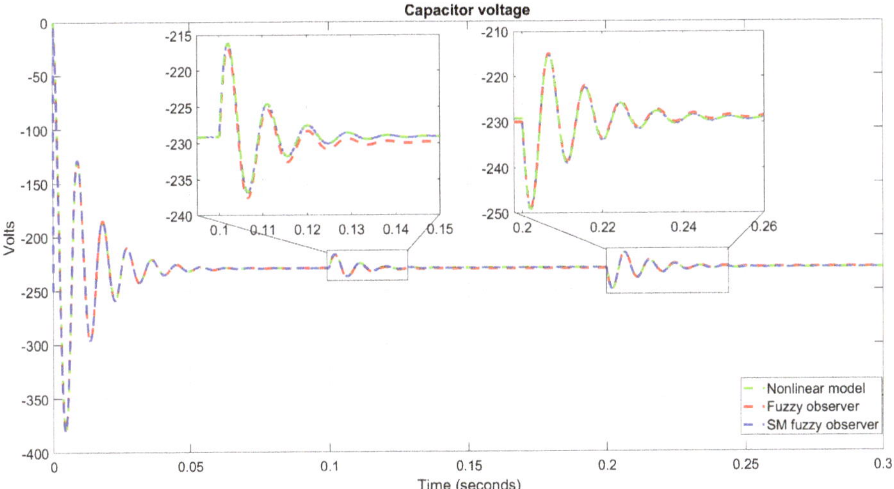

Figure 17. Observer response for the voltage of v_C with variations in load R.

Figure 18 shows the observer convergence in the inductor current i_L under variations in the nominal magnitude of the load ($R = 70.3\ \Omega$). In 0.1 s, the load is decreased to 78.23% ($R = 55\ \Omega$) and, in 0.2 s, it is increased to 113.79% ($R = 80\ \Omega$), in both cases the observer converges on the inductor current. The observer has a maximum error of 13 mA and a minimum error of 100 μA.

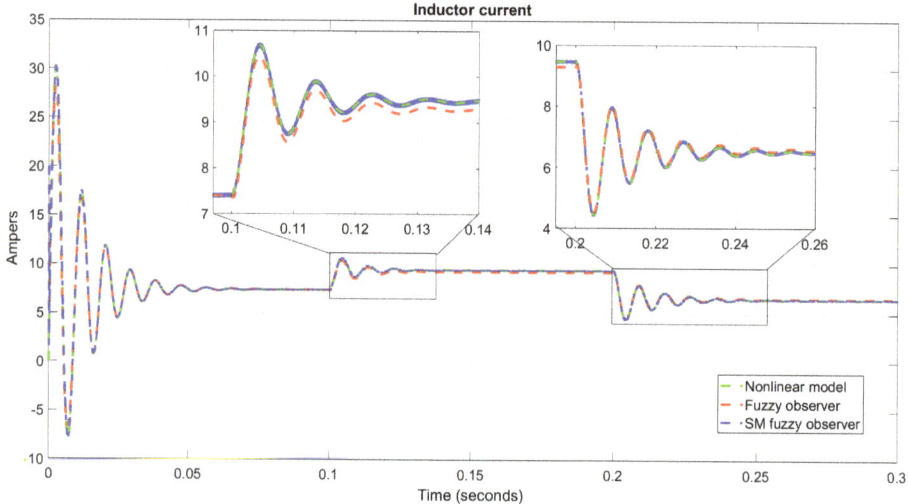

Figure 18. Observer response on inductor current i_L with variations in load R.

According to the simulation results of the fuzzy observer with sliding modes for the actuator and the experimental validation for the distillation column, it is verified that the fuzzy observer with sliding modes complies with the necessary characteristics of convergence and robustness under disturbances in order to design and implement fault detection and diagnosis systems with analytical redundancy.

6.2. Heating Behavior in the Distillation Column

Regulating the inductor current i_L and the capacitor voltage v_C in the converter implies to regulate the output electrical power, Equation (9), that energies the electrical resistor in the heating actuator of the distillation, hence regulating the heating power, as presented in Equation (5), defines the distillation dynamics.

A distillation process is performed considering an Ethanol–Water mixture in order to validate the power regulation effect in a distillation column. The parameters of the mixture components are specified in Table 3.

Table 3. Parameters of the Ethanol–Water mixture.

Parameter	Ethanol	Water
Density (ρ_c)	0.789 g/cm^3	1 g/cm^3
Molecular weight (M_{WC})	46.069 g	18.01528 g
Boiling temperature (R)	78.04 °C	78.04 °C
Vapor entalphy (ΔH^{vap})	38.56 kJ/mol·°C	40.65 kJ/mol·°C

Table 4 presents the initial parameters of the process.

Table 4. Initial parameters of the distillation process.

Parameter	Magnitude
Ethanol volume in the boiler	1000 mL
Water volume in the boiler	1000 mL
Process total pressure	636 mmHg
Sampling time	3 s

The minimum electrical power required to boil the selected mixture is experimentally determined in 180 watts for the used distillation pilot plant.

This distillation pilot plant is formed by the boiler (plate 1), nine plates, and the condenser (plate 11). Seven RTD Pt-100 sensors are located in plates 1, 2, 4, 6, 8, 10, and 11, allowing to monitor the plates temperatures required to estimate the mixture compositions in the column.

In the state-space model of a distillation column described in Equation (58), the inputs to the system are the heating power (Qb), and the reflux valve (Rf), where the heating power is defined by the electrical power used in the boiler heating actuator defined by the output voltage and the heating resistance, determined by Equation (59). The heating power influence in the thermal behavior of the column is denoted in Equation (5).

$$\dot{x} = Ax + B\begin{pmatrix} Rf \\ Qb \end{pmatrix} \tag{58}$$

$$Qb = Pt = \frac{{v_C}^2}{R}t \tag{59}$$

Two different tests are performed to validate the regulation effect of the heating power in the distillation process. In both processes, the distillation column is heated and maintained in the steady-state, when considering that the reflux valve is always off, i.e., the not distilled product is retired from the column.

Figure 19 shows the temperatures measured in the distillation plates when a non-regulated power supply is used.

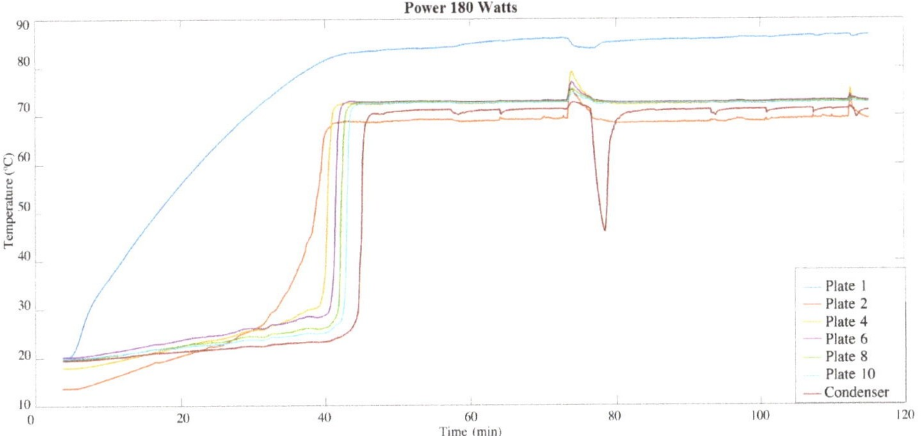

Figure 19. Thermal behavior of the distillation column using a non-regulated power supply.

Figure 20 shows the temperatures measured in the distillation plates when the presented Buck-Boost converter is used.

As can be seen in Figure 19, when the electrical power supplied to the heating actuator of the distillation column is not regulated, the thermal performance of the distillation process is unstable, which can derive in several problems, due to a thermal shock to the boiler mixture being able to cause a violent siphon effect, affecting the measurements and the rate of the distillation process. These problems are avoided when using a controlled power, as shown in Figure 20.

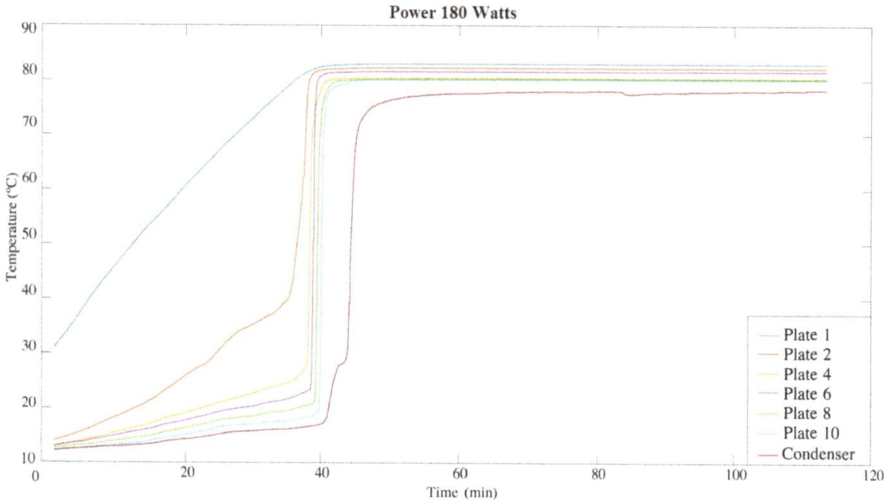

Figure 20. Thermal behavior of the distillation column using the regulated power supply.

7. Conclusions

The heating behavior of the distillation column depends mostly on the boiler and its heating actuator, which must be monitored to guarantee its adequate operation.

In this work, a fuzzy model and two fuzzy observer to estimate the inductor current and capacitor voltage in a DC-DC Buck-Boosy power converter, used to regulate the heating power in a distillation column boiler, are presented.

The observer is based on the Takagi–Sugeno fuzzy model of four rules. The gains are calculated by means of LMIs in order to guarantee the stability for each of the closed loop linear subsystems.

The fuzzy model and fuzzy observers are validated in simulation. Different tests were carried out with ideal and different initial conditions between the nonlinear system and the fuzzy observer, as well as disturbances in the nonlinear system, in order to validate the fuzzy observer convergence.

The Takagi–Sugeno fuzzy observer has an adequate response as long as the disturbances occur in the operating points selected in its designing stage. By adding the sliding-mode term, the fuzzy observer increases its robustness under load and input voltage perturbations, making it suitable to be applied in different control strategies, such as Fault Detection and Isolation (FDI) and Fault Tolerant Control (FTC) systems.

Author Contributions: Conceptualization, J.A.-M. and A.d.C.T.-A.; methodology and formal analysis, M.H.-C.; writing—review and editing, A.d.C.T.-A. and M.H.-C.; supervision, E.E.-J. All authors have read and agreed to the published version of the manuscript.

Funding: This research received funding from Scientific Research Coordination, UMSNH (Conv. 2016–2017) and Conacyt grant number 401203.

Acknowledgments: Authors thank Juan Jesús Silva Romero for the experimentation support.

Conflicts of Interest: The authors declare no conflict of interest.

References

1. Asprion, N. Modeling, Simulation, and Optimization 4.0 for a Distillation Column. *Chemie Ingenieur Technik* **2020**, *92*, 879–889. [CrossRef]
2. Albuquerque, A.A.; Ng, F.T.; Danielski, L.; Stragevitch, L. Phase equilibrium modeling in biodiesel production by reactive distillation. *Fuel* **2020**, *271*, 117688. [CrossRef]

3. Keßler, T.; Kunde, C.; McBride, K.; Mertens, N.; Michaels, D.; Sundmacher, K.; Kienle, A. Global optimization of distillation columns using explicit and implicit surrogate models. *Chem. Eng. Sci.* **2019**, *197*, 235–245. [CrossRef]
4. Piltan, F.; TayebiHaghighi, S.; Jowkar, S.; Bod, H.R.; Sahamijoo, A.; Heo, J.S. A Novel Intelligent ARX-Laguerre Distillation Column Estimation Technique. *Int. J. Intell. Syst. Appl.* **2019**, *11*, 52–60. [CrossRef]
5. Salas, S.; Romagnoli, J.A.; Tronci, S.; Baratti, R. A geometric observer design for a semi-batch free-radical polymerization system. *Comput. Chem. Eng.* **2019**, *126*, 391–402. [CrossRef]
6. Yin, X.; Liu, J. Distributed state estimation for a class of nonlinear processes based on high-gain observers. *Chem. Eng. Res. Des.* **2020**, *160*, 20–30. [CrossRef]
7. Paraschiv, N.; Olteanu, M. Feedforward process control of a distillation column based on evolutionary techniques. In Proceedings of the 2015 19th International Conference on System Theory, Control and Computing (ICSTCC), Cheile Gradistei, Romania, 14–16 October 2015; pp. 730–735.
8. Shafeeq, A.; Daood, S.S.; Muhammad, A.; Ijaz, A. Effect of variable reflux ratio on binary distillation in a laboratory scale distillation column. In Proceedings of the 2010 2nd International Conference on Chemical, Biological and Environmental Engineering, Cairo, Egypt, 2–4 November 2010; pp. 35–38.
9. Venkatramanan, D.; John, V. Dynamic modeling and analysis of buck converter based solar PV charge controller for improved MPPT performance. *IEEE Trans. Ind. Appl.* **2019**, *55*, 6234–6246. [CrossRef]
10. Spiliotis, K.; Gonçalves, J.E.; Van De Sande, W.; Ravyts, S.; Daenen, M.; Saelens, D.; Baert, K.; Driesen, J. Modeling and validation of a DC/DC power converter for building energy simulations: Application to BIPV systems. *Appl. Energy* **2019**, *240*, 646–665. [CrossRef]
11. Ortigoza, R.S.; Juarez, J.N.A.; Sanchez, J.R.G.; Cruz, M.A.; Guzman, V.M.H.; Taud, H. Modeling and experimental validation of a bidirectional DC/DC buck power electronic converter-DC motor system. *IEEE Lat. Am. Trans.* **2017**, *15*, 1043–1051. [CrossRef]
12. Fan, J.W.T.; Chow, J.P.W.; Chan, W.T.; Zhang, K.; Relekar, A.; Ho, K.W.; Tung, C.P.; Wang, K.W.; Chung, H.S.H. Modeling and experimental assessment of the EMI characteristics of switching converters with power semiconductor filters. *IEEE Trans. Power Electron.* **2019**, *35*, 2519–2533. [CrossRef]
13. Xu, S.; Zhang, J.; Huang, Y.; Jatskevich, J. Dynamic average-value modeling of three-level T-type grid-connected converter system. *IEEE J. Emerg. Sel. Top. Power Electron.* **2019**, *7*, 2428–2442. [CrossRef]
14. Dong, X.; Griffo, A.; Hewitt, D.A.; Wang, J. Reduced-Order Thermal Observer for Power Modules Temperature Estimation. *IEEE Trans. Ind. Electron.* **2019**, *67*, 10085–10094. [CrossRef]
15. Grégoire, L.A.; Wang, W.; Seleme, S.I.; Fadel, M. High reliability observers for modular multilevel converter capacitor voltage evaluation. In Proceedings of the 2016 IEEE 8th International Power Electronics and Motion Control Conference (IPEMC-ECCE Asia), Hefei, China, 22–26 May 2016; pp. 2332–2336.
16. Zhuo, S.; Gaillard, A.; Xu, L.; Liu, C.; Paire, D.; Gao, F. An Observer-Based Switch Open-Circuit Fault Diagnosis of DC-DC Converter for Fuel Cell Application. *IEEE Trans. Ind. Appl.* **2020**, *56*, 3159–3167. [CrossRef]
17. Li, J.; Zhang, Z.; Li, B. Sensor fault detection and system reconfiguration for DC-DC boost converter. *Sensors* **2018**, *18*, 1375. [CrossRef] [PubMed]
18. Tanaka, K.; Sugeno, M. Stability analysis and design of fuzzy control systems. *Fuzzy Sets Syst.* **1992**, *45*, 135–156. [CrossRef]
19. Skogestad, S. Dynamics and control of distillation columns: A tutorial introduction. *Chem. Eng. Res. Des.* **1997**, *75*, 539–562. [CrossRef]
20. Hyde, J.; Cuspinera, A.; Regué, J. *Control Electroneumático y Electrónico*; Marcombo: Barcelona, Spain, 1997; Volume 2.
21. Rashid, M.H. *Power Electronics Handbook*; Butterworth-Heinemann: Oxford, UK, 2017.
22. Boora, A.A.; Zare, F.; Ledwich, G.; Ghosh, A. A general approach to control a positive buck-boost converter to achieve robustness against input voltage fluctuations and load changes. In Proceedings of the 2008 IEEE Power Electronics Specialists Conference, Rhodes, Greece, 15–19 June 2008; pp. 2011–2017.
23. Zhou, X.; He, Q. Modeling and simulation of Buck-Boost converter with voltage feedback control. In Proceedings of the 2015 7th International Conference on Mechanical and Electronics Engineering, Bali, Indonesia, 1–2 July 2015.

24. Çorapsiz, M.R.; Kahveci, H. Voltage Control Strategy for DC-DC Buck-Boost Converter. In Proceedings of the 4th International Conference on Advanves in Natural & Applied Sciences, Agri, Turkey, 19–22 June 2019; pp. 436–446.
25. Feshara, H.F.; Ibrahim, A.M.; El-Amary, N.H.; Sharaf, S.M. Performance evaluation of variable structure controller based on sliding mode technique for a grid-connected solar network. *IEEE Access* **2019**, *7*, 84349–84359. [CrossRef]
26. Javed, K.; Vandevelde, L.; De Belie, F. Feed-Forward Control Method for Digital Power Factor Correction in Parallel Connected Buck-Boost Converter (CCM Mode). In Proceedings of the 2020 International Symposium on Power Electronics, Electrical Drives, Automation and Motion (SPEEDAM), Sorrento, Italy, 24–26 June 2020; pp. 827–831.
27. Arazi, M.; Payman, A.; Camara, M.B.; Dakyo, B. Study of different topologies of DC-DC resonant converters for renewable energy applications. In Proceedings of the 2018 Thirteenth International Conference on Ecological Vehicles and Renewable Energies (EVER), Monte Carlo, Monaco, 10–12 April 2018; pp. 1–6.
28. Rojas, C.A.; Kouro, S.; Perez, M.A.; Echeverria, J. DC-DC MMC for HVDC grid interface of utility-scale photovoltaic conversion systems. *IEEE Trans. Ind. Electron.* **2018**, *65*, 352–362. [CrossRef]
29. Tanaka, K.; Ikeda, T.; Wang, H.O. Fuzzy regulators and fuzzy observers: Relaxed stability conditions and LMI-based designs. *IEEE Trans. Fuzzy Syst.* **1998**, *6*, 250–265. [CrossRef]
30. Anzurez-Marín, J. Diagnóstico de Fallas en Sistemas no Lineales Usando LóGica Difusa y Observadores con Modos Deslizantes. Ph.D. Thesis, CINVESTAV, Mexico City, Mexico, 2007.
31. Castillo-Toledo, B.; Anzurez-Marin, J. Model-based fault diagnosis using sliding mode observers to Takagi–Sugeno fuzzy model. In Proceedings of the 2005 IEEE International Symposium on Mediterrean Conference on Control and Automation Intelligent Control, Limassol, Cyprus, 27–29 June 2005; pp. 652–657.
32. Luenberger, D. Observers for multivariable systems. *IEEE Trans. Autom. Control* **1966**, *11*, 190–197. [CrossRef]
33. Bergsten, P.; Palm, R.; Driankov, D. Observers for Takagi–Sugeno fuzzy systems. *IEEE Trans. Syst. Man Cybern. B Cybern.* **2002**, *32*, 114–121. [CrossRef] [PubMed]

© 2020 by the authors. Licensee MDPI, Basel, Switzerland. This article is an open access article distributed under the terms and conditions of the Creative Commons Attribution (CC BY) license (http://creativecommons.org/licenses/by/4.0/).

Article

Simultaneous Optimal Estimation of Roughness and Minor Loss Coefficients in a Pipeline

Ildeberto Santos-Ruiz [1,2,*], Francisco-Ronay López-Estrada [1], Vicenç Puig [2] and Guillermo Valencia-Palomo [3]

1. TURIX-Dynamics Diagnosis and Control Group, Tecnológico Nacional de México, IT Tuxtla Gutiérrez, Carretera Panamericana Km 1080, Tuxtla Gutiérrez, Chiapas C.P. 29050, Mexico; frlopez@ittg.edu.mx
2. Institut de Robòtica i Informàtica Industrial, CSIC-UPC, Universitat Politècnica de Catalunya, C/. Llorens i Artigas 4-6, 08028 Barcelona, Spain; vicenc.puig@upc.edu
3. Tecnológico Nacional de México, IT Hermosillo, Ave. Tecnológico y Periférico Poniente S/N, Hermosillo, Sonora C.P. 83170, Mexico; gvalencia@hermosillo.tecnm.mx

* Correspondence: idelossantos@ittg.edu.mx; Tel.: +52-961-615-0461

Received: 7 August 2020; Accepted: 31 August 2020; Published: 1 September 2020

Abstract: This paper presents a proposal to estimate simultaneously, through nonlinear optimization, the roughness and head loss coefficients in a non-straight pipeline. With the proposed technique, the calculation of friction is optimized by minimizing the fitting error in the Colebrook–White equation for an operating interval of the pipeline from the flow and pressure measurements at the pipe ends. The proposed method has been implemented in MATLAB and validated in a serpentine-shaped experimental pipeline by contrasting the theoretical friction for the estimated coefficients obtained from the Darcy–Weisbach equation for a set of steady-state measurements.

Keywords: nonlinear optimization; turbulent flow; friction factor; pipe roughness; minor losses

1. Introduction

Pipelines are one of the most economical means of transporting liquids. They are widely used to carry water, fuels, and other substances in the process industries. Pipeline monitoring generally consists of measuring pressures and flow rates at some strategic points in the pipeline, frequently including the ends of the line. The changes in the hydraulic variables, mainly the pressures, make it possible to identify disturbances and faults in the system, e.g., blockages and leaks, as well as the implementation of control loops that allow regulating or changing the operating point [1]. Since many of the diagnosis/control algorithms are based on a mathematical model of the hydraulic system, fine-tuning such algorithms requires precise knowledge of their parameters [2,3]. For example, to diagnose leaks through pressure variations, it is essential to know how much pressure variation is caused by a reasonable pressure drop due to friction in the pipeline, which is only possible if the roughness and head loss coefficients of the pipe are known [4,5].

The Colebrook–White equation is the primary reference for calculating the friction factor in pressurized pipes. Since there are no known explicit solutions to this nonlinear equation in terms of elementary functions, many works have focused on studying its numerical solution [6,7] or by proposing different explicit approaches with varying degrees of accuracy [8–12].

Since many pipelines operate with turbulent flow, most publications focus on the study of friction for this regime. However, there are some works in which explicit formulas have been proposed to calculate the friction factor for the three regimes: laminar, transitional, and turbulent, considering both smooth and rough pipes [13]. There are also works where only friction in smooth pipes is studied, neglecting the roughness [14], and others where pipes with ribs are considered [15].

Then, by considering the importance of friction in head losses, it could be said that the reports on the investigation of physical measurements and mathematical modeling of pipe roughness are scarce in the technical literature [16]. Direct measurement of roughness is not a simple process. However, current developments in profilometry and surface engineering have achieved accurate measures of a pipe's internal roughness, but only in a laboratory setting, usually with new pipes. In this way, Farshad and Pesecreta evaluated the roughness of different pipes' rough surfaces using linear surface profilometers [17]. However, Kang and Lansey point out that, in practice, the roughness of the pipe cannot be measured in the field, so it must be estimated indirectly [18]. In this context, the research in [19] addresses the estimation of roughness in a straight water supply pipe from flow measurements and head losses measured by piezometric pipes. Since the experimental pipe they used in this work is approximately straight, the minor head losses due to the fittings were not considered.

The works above focus on calculating the friction factor, which is the critical element in determining head losses in long-distance pipelines. Fittings and valves contribute very little to the total pressure drop in the pipeline. Therefore, in such cases, pressure losses through fittings, valves, and other restrictions are generally classified as "minor losses" and are neglected. However, in shorter pipelines, the pressure loss in valves, fittings, etc., can be a substantial part of the total pressure loss, so these must be included for correct engineering calculations [16]. Therefore, it is necessary to know the head loss coefficients (for local losses, accessories, and valves), and the roughness coefficients related to friction (for distributed losses, in the entire pipeline).

During the design of the piping system, it may be sufficient to know the nominal value of the roughness and head loss coefficients provided by the manufacturer. The calculations consider a safety factor that guarantees the operation of the system with an oversized pumping power [17]. However, during system operation, the implementation of model-based monitoring and diagnosis algorithms require an accurate estimate of both roughness and head loss coefficients. It should be considered that, even under the same pumping pressure, the head losses change over time because, with continuous use, the pipelines show natural wear and aging. Particles can also accumulate on the pipe walls, changing the roughness and effective diameter of the pipes [18,19]. Therefore, a periodic recalibration is necessary to update the parameter values.

Recently, in [20], both major losses (due to roughness) and minor losses (due to fittings) are independently studied. However, as will be explained in Section 3, when these are calculated separately, a small overestimate in one of them leads to an overestimate in the other one, and vice versa. This article proposes a procedure to simultaneously estimate the roughness and minor loss coefficients in a pipeline based on these considerations. The proposed calculation requires some measurements of flow and pressure at the pipeline ends.

This work's main contribution is the simultaneous computation of the coefficients of both components of the pressure loss: the major losses associated with roughness and the minor losses related to the fittings in the pipeline. In this proposal, it is possible to jointly estimate both coefficients assuming the minor losses as equivalent pipe lengths and looking for the coefficients that minimize the fitting error in the Colebrook–White equation. In order to demonstrate the applicability of the proposed technique, a series of tests on a prototype pipeline is described.

2. Background on Turbulent Flow in Pipelines

There is a general assumption that the head loss (h_f) due to friction depends on the inner diameter (D) of the pipe, the length (L) in which the head loss is measured, the average flow velocity (V), the absolute roughness of the pipe wall (k_s), the gravity acceleration (g), and the density and the viscosity of the fluid. Through force balance and dimensional analysis, it is possible to determine the head losses due to friction as [21,22]:

$$h_f = \frac{L}{D}\frac{V^2}{2g}f(\varepsilon, \mathrm{Re}), \tag{1}$$

where Re is the Reynolds number, which measures flow turbulence as a function of viscosity and velocity, and $\varepsilon = k_s/D$ is the so-called relative roughness coefficient. The expression $f(\varepsilon, \text{Re})$ in (1) is the "friction factor", and is abbreviated hereinafter with the shortcut $f := f(\varepsilon, \text{Re})$. In this way, Equation (1) is represented in the following simplified form called the Darcy–Weisbach equation [21,23]:

$$h_f = f\frac{L}{D}\frac{V^2}{2g} = f\frac{8L}{g\pi^2 D^5}Q^2, \tag{2}$$

where Q is the volumetric flow rate in the pipeline. Compared to other formulas for calculating friction losses, Equation (2) has the advantage of being a dimensionally homogeneous equation, so that the friction factor f is a dimensionless number consistent with any system of units [21]. There are works where some variants of the Darcy–Weisbach equation are considered. For example, [24] suggests an improvement in the calculation of the friction factor by assuming a periodicity of the roughness in the longitudinal direction of the pipe. However, in this work only the classical form (2) is considered.

In general, the friction factor depends on the flow turbulence (measured with Re) and also on the physical characteristics of the pipe (diameter and roughness of the interior walls). However, when the flow is laminar, at low velocity, the friction factor depends only on the Reynolds number, and is given by $f = 64/\text{Re}$ for $\text{Re} \leq 2000$. Instead, at high speeds where the flow is turbulent ($\text{Re} \geq 4000$), the friction factor also depends on the roughness of the pipe walls. In the interval, $2000 < \text{Re} < 4000$, after the laminar regime and before the turbulent regime, the values of the friction factor are not very predictable. Therefore this transitional regime will not be considered in this work.

One of the most widely accepted models for expressing the relationship between friction, turbulence, and roughness in pressurized flows is the Colebrook–White equation:

$$\frac{1}{\sqrt{f}} + 2\log_{10}\left(\frac{\varepsilon}{3.7} + \frac{2.51}{\text{Re}\sqrt{f}}\right) = 0, \tag{3}$$

where the Reynolds number given by

$$\text{Re} = \frac{DQ}{A\nu} \tag{4}$$

measures flow turbulence in terms of kinematic viscosity ν and flow rate Q, while the friction factor

$$f = \frac{g\pi^2 D^5 h_f}{8LQ^2} \tag{5}$$

is expressed in terms of the head loss h_f $(= H_{\text{in}} - H_{\text{out}})$ along the pipe.

Explicit solutions for f of the DW equation using elementary functions are not known, because due to their nonlinearity, it is impossible to isolate f in terms of ε and Re. However, it is possible to find approximate solutions using iterative numerical methods. For this, the DW equation is rewritten as:

$$\phi(f) := \frac{1}{\sqrt{f}} + 2\log_{10}\left(\frac{\varepsilon}{3.7} + \frac{2.51}{\text{Re}\sqrt{f}}\right) = 0. \tag{6}$$

The problem is then reduced to iteratively find the zero of $\phi(f)$ or, equivalently, the minimum of $\phi^2(f)$. Figure 1 shows the numerical solution of (6) obtained using the COLEBROOK MATLAB-routine, which can be consulted in [20]. It is also possible to estimate the solution of the CW equation, with different percentages of error, using explicit non-iterative approximations, such as those proposed by [25–27], among others. The surface plot in Figure 1 shows the general behavior that friction factor increases with roughness and decreases with the Reynolds number. However, for high turbulence (very large Re) the friction factor depends only on the roughness.

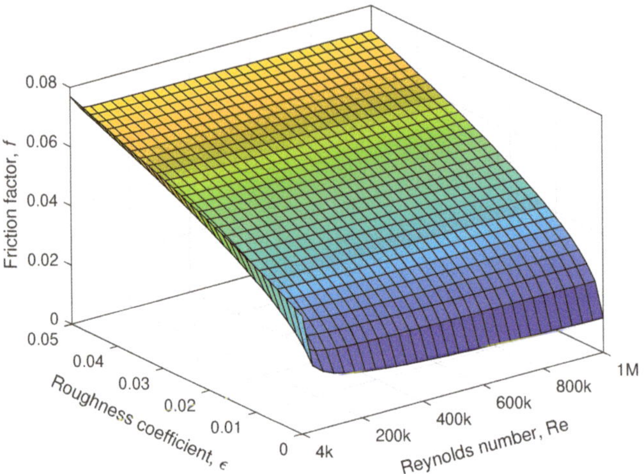

Figure 1. Variation of friction factor with roughness and Reynolds number in turbulent regime.

On the other hand, in addition to the pressure losses due to friction, which are distributed throughout the pipeline, the pressure losses due to flow disturbance caused by elbows, valves, and other fittings must be quantified. These are not distributed losses, but rather local losses that can be located in a well-defined position. Head losses due to singularities or fittings are commonly called "minor losses" because, for very long pipes, their value can be neglected compared to that due to friction. However, their effect can be significant for short pipes [28].

Equation (2) only predicts major losses due to fluid friction on the pipe wall and due to the effects of fluid viscosity and does not include minor losses on inputs, elbows, and other fittings. Minor losses in accessories are defined in terms of a loss coefficient K, by

$$h_m = K\frac{V^2}{2g}, \tag{7}$$

which has some similarity with Equation (2) used for friction losses. There are some empirical formulas to estimate the minor loss coefficient (K) for valves, elbows, and other fittings [16]. In practice, the values obtained in this way can differ considerably from the true values, so it is convenient to determine them experimentally. For example, for 90° elbows, the minor loss coefficient is typically in the interval $0.9 \leq K \leq 1.5$, according to [29]; on the other hand, in [16] it is suggested to use the empirical formula $K = 30 f_T$, where f_T is the turbulent friction factor. In general, the minor loss coefficient is a function of both the fitting geometry and the Reynolds number. However, when Re is large enough, K can be assumed as a function of the fitting geometry only.

Given the similarity between Equations (2) and (7), in some applications it has been proposed to combine distributed losses with local losses, expressing the latter as if they were friction losses in an additional length of pipe. The "equivalent length" of pipe associated with a local head loss is the length of pipe that would produce a friction head loss equal to the corresponding local head loss. By matching Equations (2) and (7), the equivalent length L_{eq} can be obtained as

$$K\frac{V^2}{2g} = f\frac{L_{\text{eq}}}{D}\frac{V^2}{2g} \quad \Rightarrow \quad L_{\text{eq}} = \frac{KD}{f}, \tag{8}$$

so, if the additional length of pipe that would equal the local losses is known, the minor loss coefficient is given by

$$K = f\frac{L_{\text{eq}}}{D}. \tag{9}$$

The above equations are the basis for addressing the problem under study. Formally, the problem whose solution is proposed in the next section can be expressed as follows: "Given a list of flow-rate and pressure measurements at pipeline ends, find simultaneously the roughness coefficient ε and the minor loss coefficient K—as equivalent length—so that for the friction factor values obtained experimentally, via the Darcy–Weisbach equation, the fitting error in the Colebrook–White equation is minimized".

3. Proposed Methodology

Consider the simultaneous estimation of the roughness coefficient and the minor loss coefficient. The strategy used to integrate the minor losses together with the friction (major losses) is to assume them as an additional length of pipe, so that the total length used in the calculations is a "computational length" that integrates both the physical length of the pipeline as the equivalent length of the fittings.

If the pipe length L is assumed to be known, the friction factor f can be determined by measuring the flow rate Q and the head loss h_f between the pipeline ends and substituting these values (5). Then, the roughness can be obtained by substituting f in the nonlinear Equation (3) and solving for ε. However, when starting from an underestimated length, if the roughness is overestimated, then the pipeline appears to be rougher when its length is shortened for a given friction factor.

Additionally, in practice, the roughness ε obtained by the numerical solution of (3) is sensitive to variations in the computed value of f due to measurement noise on Q and h_f. Therefore, a single value of ε does not completely satisfy (3) for different measurements, even if they correspond to the same operating point. Consequently, for each measurement, the right side of (3) is not exactly zero, but there is a residual or fitting error δ. However, if we consider a length-dependent friction factor, $f(L)$, it is possible to minimize the error δ by adjusting the parameters ε and L in (3). In this way, the Colebrook–White equation expresses the friction factor in terms of flow rate and head loss but is parameterized by the roughness coefficient and the total length of the pipeline.

Thus, for each value of flow rate and head loss (for each operating point), there is a fitting error

$$\delta_k(\varepsilon, L) := \frac{1}{\sqrt{f_k(L)}} + 2\log_{10}\left(\frac{\varepsilon}{3.7} + \frac{2.51}{\mathrm{Re}_k\sqrt{f_k(L)}}\right), \tag{10}$$

where Re_k and f_k are calculated as specified in (4) and (5), using the flow rate and head loss measurements in the k-th operating point.

Considering a nonlinear least squares optimization problem, with N measurements ($N \geq 2$), the optimal roughness and total length estimates have been defined as follows:

$$\widehat{\varepsilon}, \widehat{L} := \underset{\varepsilon, L}{\arg\min} \sum_{k=1}^{N} \delta_k(\varepsilon, L)^2 = \underset{\varepsilon, L}{\arg\min}\, \delta^\top \delta, \tag{11}$$

where $\delta = [\delta_1(\varepsilon, L), \delta_2(\varepsilon, L), \ldots, \delta_N(\varepsilon, L)]^\top$ is the vector of fitting errors in the Colebrook–White equation. One way to solve the optimization problem (11), the one proposed in this work, is through the Levenberg–Marquardt (LM) method [30,31].

In the context of curve fitting, the LM method, also known as the damped least squares (DLS) method, is formulated as follows: Given a set of N pairs of empirical data $(\mathbf{x}_k, y_k)$ of independent and dependent variables, find the parameters $\boldsymbol{\beta}$ of the model $\phi(\mathbf{x}, \boldsymbol{\beta})$ so that the sum of the squares of the deviations is minimized:

$$\widehat{\boldsymbol{\beta}} = \underset{\boldsymbol{\beta}}{\arg\min} \sum_{k=1}^{N} [y_k - \phi(\mathbf{x}_k, \boldsymbol{\beta})]^2. \tag{12}$$

There is a correspondence between the optimization form (12) and the problem of calculating the roughness and length in (11), assuming the following equivalence:

$$\boldsymbol{\beta} \equiv [\varepsilon, L]^{\top}, \quad y_k \equiv 0, \quad \phi(\mathbf{x}_k, \boldsymbol{\beta}) \equiv \delta_k(\varepsilon, L). \tag{13}$$

In order to find the optimal value of $\boldsymbol{\beta}$, the LM algorithm iterates from an initial approximation $\boldsymbol{\beta}_0$. In each iteration, the parameter vector, $\boldsymbol{\beta}$, is replaced by a new estimate, $\boldsymbol{\beta} + \boldsymbol{\Delta}$. The increment $\boldsymbol{\Delta}$ is calculated, according to [32], by solving

$$\left(\mathbf{J}^{\top}\mathbf{J} + \lambda \operatorname{diag}\left(\mathbf{J}^{\top}\mathbf{J}\right)\right) \boldsymbol{\Delta} = \mathbf{J}^{\top} \left(\mathbf{y} - \boldsymbol{\phi}(\boldsymbol{\beta})\right), \tag{14}$$

where $\mathbf{J} = [\partial\phi(\mathbf{x}_k, \boldsymbol{\beta})/\partial\boldsymbol{\beta}]$ is the Jacobian matrix, $\boldsymbol{\phi}(\boldsymbol{\beta}) = [\phi(\mathbf{x}_k, \boldsymbol{\beta})]$, $\mathbf{y} = [y_k]$, and λ is a damping factor updated at each iteration. The iterations start with an empirical value $\lambda = 0.01$. Then if the sum in (12) decreases fast, a smaller value $\lambda \leftarrow \lambda/10$ is used in the subsequent iteration. When the step size, $\boldsymbol{\Delta}$, or the decrease in the sum of squares for the last parameter vector, $\boldsymbol{\beta} + \boldsymbol{\Delta}$, falls below predefined limits, the iteration stops, and the last parameter vector $\boldsymbol{\beta}$ is considered the solution. The box in Figure 2 summarizes the complete calculation procedure.

Input. Column-shaped arrays of the same length: H_{in}, H_{out}, Q, and ν.

1. Compute Re and f using Equations (4) and (5).
2. Define the cost function $C(\boldsymbol{\beta}) := \boldsymbol{\phi}(\boldsymbol{\beta})^{\top}\boldsymbol{\phi}(\boldsymbol{\beta})$, taking $\boldsymbol{\beta} = [\varepsilon, L]^{\top}$ and $\boldsymbol{\phi}(\boldsymbol{\beta}) = [\delta_k(\varepsilon, L)]$ according to Equation (10).
3. Define an initial approximation $\boldsymbol{\beta}_0$, a damping factor λ (suggested, 0.01), and tolerances TOL_{Δ} (suggested, 1×10^{-8}) and TOL_{β} (suggested, 1×10^{-8}).
4. $\boldsymbol{\beta} \leftarrow \boldsymbol{\beta}_0$
5. **repeat**
   (a) Compute the Jacobian matrix $\mathbf{J}$.
   (b) Solve Equation (14) for $\boldsymbol{\Delta}$.
   (c) $\boldsymbol{\beta}_{\text{old}} \leftarrow \boldsymbol{\beta}$
   (d) $\boldsymbol{\beta} \leftarrow \boldsymbol{\beta} + \boldsymbol{\Delta}$
   (e) **if** $C(\boldsymbol{\beta}) < C(\boldsymbol{\beta}_{\text{old}})$ **then**
      $\lambda \leftarrow \lambda/10$
   else
      $\lambda \leftarrow 10\lambda$
   endif

   until $\|\boldsymbol{\Delta}\| \le \mathsf{TOL}_{\Delta}$ or $\|\boldsymbol{\beta} - \boldsymbol{\beta}_{\text{old}}\| \le \mathsf{TOL}_{\beta}$

Output. Optimal estimates: $\widehat{\varepsilon} = \boldsymbol{\beta}[1]$, $\widehat{L} = \boldsymbol{\beta}[2]$.

Figure 2. Algorithm to compute the optimal estimates of the roughness and minor loss coefficients.

The solution to the problem (11), using the LM method as described in Figure 2, was coded into a MATLAB subroutine. This subroutine, named Pipeline Parameter Calibration (PPC), receives inlet and outlet pressures, flow rate, and viscosity measurements at each operating point of the pipeline, and returns the estimated optimal values for ε and L. The head loss coefficient K can be obtained from the excess length of $\widehat{L}$ over the straight length using (9). In slightly turbulent flows, measurements should be taken over a small operating interval, because the minor loss coefficient could vary significantly between widely separated operating points. However, for completely turbulent flows the minor loss coefficient remains approximately constant. The range of Re in the test measurements described in the next section corresponds to a sufficiently high turbulence.

4. Experimental Setup

The procedure described in the preceding section has been tested on a prototype serpentine-shaped pipeline [33] (see Figure 3). This experimental pipeline is part of the Hydroinformatics Laboratory of the National Institute of Technology of Mexico, located in Tuxtla Gutiérrez. This pipeline has a straight length of 84.58 m and uses 90° elbows to give it a serpentine shape, as shown in Figure 4. The flow is driven from a 2500 l tank using a 5 hp centrifugal pump whose power is controlled by a frequency inverter. Pressure and flow sensors/transmitters are available at the pipeline ends. Although the prototype pipeline has two flow sensors, only one of them is used to obtain the flow-rate measurements required by the algorithm, because the measurements are made in a steady state with no leaks. There are no pressure sensors before and after each change in flow direction, so it is not possible to calculate the head loss on each direction change individually. In order to change the operating point of the pipeline, the frequency inverter was driven at the following working frequencies: 30 Hz, 35 Hz, 40 Hz, 45 Hz, 50 Hz and 55 Hz. Other parameters of the experimental pipeline used in calculations are $D = 0.0486$ m, and $g = 9.79\,\mathrm{m/s^2}$.

Figure 3. Prototype pipeline in the National Institute of Technology of Mexico, at Tuxtla Gutiérrez.

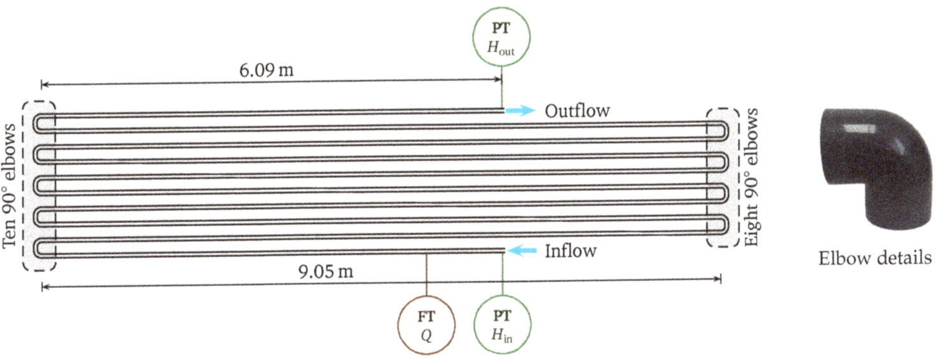

Figure 4. Top view of the prototype pipeline.

Table 1 shows the first dataset containing the pressure and flow measurements used to estimate the roughness and equivalent length of the prototype pipe. The kinematic viscosity in the fourth column was calculated from the measured water temperature by spline interpolation using the NIST Standard Reference Data, according to [34]. Pressure and flow measurements at each operating point were obtained by averaging the sensor signals over a time interval (1000 samples, corresponding to 10 s) to minimize the effect of measurement noise. In order to test the Algorithm of Figure 2, five datasets with the same structure as Table 1 were built, named `pipeline_data_1:pipeline_data_5`. Each dataset contains measurements at six operating points within the typical working range in the prototype pipeline, limited by the available pumping power. In practice, in other pipelines, the measurements used should consider their specific operating region, provided that it corresponds to the turbulent regime because the Colebrook–White equation does not correctly represent the laminar and transitional regimes.

Table 1. Dataset with measurements at six operating points.

H_{in} [m]	H_{out} [m]	Q [m^3/s]	ν [m^2/s]
3.7528	1.6063	0.0016903	8.4116×10^{-7}
4.7461	1.7960	0.0020166	8.3951×10^{-7}
5.8550	2.0033	0.0023346	8.3713×10^{-7}
7.0908	2.2399	0.0026452	8.3513×10^{-7}
8.4442	2.4941	0.0029531	8.3296×10^{-7}
9.8962	2.7663	0.0032562	8.3061×10^{-7}

5. Results and Discussion

The box in Figure 5 shows the result of running the PPC subroutine with the measurements reported in Table 1. The best fit of the measurements to the Colebrook–White equation was obtained for a roughness coefficient $\widehat{\varepsilon} = 3.4652 \times 10^{-4}$ and a total length $\widehat{L} = 112.2238$ m. This result corresponds to an excess length of 27.6438 m, compared to the physical length, due to minor losses. Figure 6 graphically shows the existence of the minimum, evidencing the behavior of the cost function in the neighborhood of the optimal estimate $\widehat{\beta} = (\widehat{\varepsilon}, \widehat{L})$. Starting from the initial approximation $\beta_0 = (1 \times 10^{-4}, 100)$, the LM procedure converges to the optimal value in eight iterations, when the norm of the increment, $\|\Delta\|$, reaches the value of 5.1559×10^{-9}.

```
>> load pipeline_data_1
>> ppc(Hin, Hout, Q, nu)

Dataset...
   Hin        Hout         Q                nu              Re          f
  ------     ------     ---------       ----------       ------     --------

  3.7528     1.6063     0.0016903       8.4116e-07        52644     0.021924
  4.7461     1.7960     0.0020166       8.3951e-07        62932     0.021168
  5.8550     2.0033     0.0023346       8.3713e-07        73063     0.020620
  7.0908     2.2399     0.0026452       8.3513e-07        82980     0.020231
  8.4442     2.4941     0.0029531       8.3296e-07        92881     0.019909
  9.8962     2.7663     0.0032562       8.3061e-07       102704     0.019622

Calibration results...
epsilon~=~0.00034652
L~=~112.2238
```

Figure 5. MATLAB output of the Pipeline Parameter Calibration (PPC) subroutine for the measurements in Table 1.

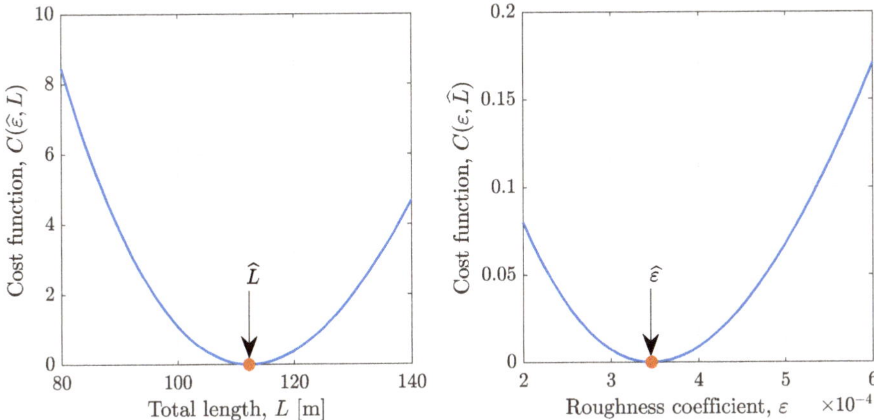

Figure 6. Behavior of the cost function in the neighborhood of the optimal estimate.

The PPC subroutine was run in each of the five available datasets; the results are summarized in Table 2. It can be seen that the roughness estimate was more sensitive than the length estimate concerning the change in pressure and flow-rate measurements. This is evidenced by the fact that the roughness varies from the third significant digit between different datasets, while the length only varies from the fourth significant digit. More precisely, the root-mean-square deviation (RMSD) was 1.929×10^{-6} (0.56%) for roughness, and 0.0456 m (0.04%) for length.

Table 2. Roughness and length estimates from different datasets.

Dataset	$\widehat{\varepsilon}$	$\widehat{L}$ [m]
pipeline_data_1	0.00034652	112.2238
pipeline_data_2	0.00034982	112.1517
pipeline_data_3	0.00034473	112.2626
pipeline_data_4	0.00034662	112.2395
pipeline_data_5	0.00034946	112.1530

The variation in the results obtained with different datasets could be attributed, at least partially, to the measurement error of the sensors, which is between 0.01% and 0.04% (maximum) according to the manufacturer's datasheets. This has the consequence that the computation of the friction factor, based on an inaccurate estimate of the roughness coefficient, is also inaccurate. Sensitivity analysis in the working region of the prototype pipeline showed that

$$(\Delta f)/f = 0.092\,(\Delta\varepsilon)/\varepsilon, \qquad (\Delta f)/f = -0.156\,(\Delta \mathrm{Re})/\mathrm{Re}, \tag{15}$$

where $(\Delta\mathrm{Re})/\mathrm{Re}$ can be taken approximately equal to the flow-rate measurement error, $(\Delta Q)/Q$, considering the relationship (4). These results show that the friction factor calculation is considerably sensitive to flow measurement errors.

The final estimates of ε and L were obtained by averaging the estimates from the five experiments on the prototype pipeline, resulting in $\widehat{\varepsilon} = 3.4743 \times 10^{-4}$ and $\widehat{L} = 112.21$ m. On the Moody chart in Figure 7, the interval of the friction factor in the prototype pipeline has been represented in red, according to the calculated roughness coefficient. This can be considered as the "working region" of the pipeline. The solid orange line in Figure 7 is the theoretical curve according to the estimated roughness $\widehat{\varepsilon}$, while the discrete red points represent the specific values of Re and f obtained from the measurements (via Equations (4) and (5), respectively), confirming a good parameterization of the Colebrook–White equation to fit the data.

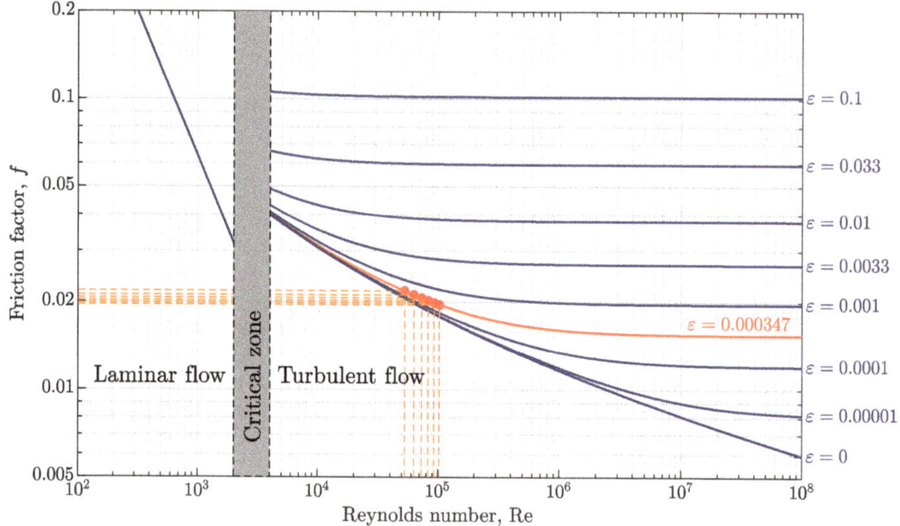

Figure 7. Moody chart with the parameters computed in the prototype pipeline.

In the prototype pipeline, minor losses are primarily due to the elbows used to give it the serpentine shape, so that the equivalent length of each elbow is determined by

$$L_{\text{elbow}} = (L_{\text{total}} - L_{\text{physical}})/n = (112.21\,\text{m} - 84.58\,\text{m})\,/18 = 1.535\,\text{m}, \tag{16}$$

where n is the number of elbows, and L_{physical} is the true length, also termed "straight length" of the pipeline.

Finally, the minor loss coefficient for each elbow can be obtained from (9). As noted in Section 2, the K values change slightly with the operating point. For example, for the fifth operating point in Table 1 (corresponding to Re $= 92881$) it was obtained:

$$K_{\text{elbow}} = f\frac{L_{\text{elbow}}}{D} = 0.629. \tag{17}$$

These K values obtained for the elbows of the prototype pipeline are close to those obtained with the empirical formula in [16] and fall outside the interval suggested by [29]. Due to the proximity between each elbows pair, it might be desirable to assign combined loss coefficients in pairs, since possibly the energy loss at the inlet elbow is different from the energy loss at the outlet elbow. In this case, the combined loss coefficient for each flow return, K_{return}, is twice that calculated by (17), since the number of elbows doubles the number of return bends.

6. Conclusions

A technique for simultaneously estimating the roughness and minor loss coefficients in non-straight pipelines has been presented and tested. The results show that, within the accuracy conditioned by the sensors' measurement uncertainty, the proposed methodology produces a reasonable estimate of both coefficients. In this regard, the roughness coefficient estimate is more sensitive to measurement noise than the minor loss coefficient estimate by almost one order of magnitude.

In support of this calculation proposal, it should be mentioned that the described technique to compute the roughness and the equivalent length has been used in multiple tests to locate leaks using extended Kalman filters, reaching a location accuracy of close to 2% of the pipeline length. The leak location procedure is not described here but can be found published in [5].

The main limitation of the proposed technique is that it only allows the estimation of minor loss coefficients for a single type of fitting. When the pipeline presents considerable head losses in more than one type of fitting, it does not determine what fraction of the minor loss is attributed to each one. As future work, it is intended to modify or extend the proposed technique to allow the calculation of roughness and minor loss coefficients in branched pipelines, also considering valves and other fittings such as tees and wyes.

Author Contributions: Conceptualization, I.S.-R., F.-R.L.-E. and V.P.; methodology, I.S.-R.; software, I.S.-R.; validation, F.-R.L.-E., V.P. and G.V.-P.; formal analysis, G.V.-P.; data curation, I.S.-R.; writing—original draft preparation, I.S.-R.; writing—review and editing, F.-R.L.-E., V.P. and G.V.-P.; supervision, F.-R.L.-E. and V.P.; project administration, F.-R.L.-E. All authors have read and agreed to the published version of the manuscript.

Funding: This research was funded by the National Council for Science and Technology (CONACYT) of Mexico, under the program Atención a Problemas Nacionales, grant number PN-2016/3595.

Conflicts of Interest: The authors declare no conflict of interest. The funders had no role in study design, data collection and analysis, decision to publish, or preparation of the manuscript.

References

1. Timashev, S.; Bushinskaya, A. *Diagnostics and Reliability of Pipeline Systems*; Springer: New York, NY, USA, 2016.
2. Puig, V.; Ocampo-Martínez, C.; Pérez, R.; Cembrano, G.; Quevedo, J.; Escobet, T. *Real-Time Monitoring and Operational Control of Drinking-Water Systems*; Springer International Publishing: New York, NY, USA, 2017.
3. Verde, C.; Torres, L. *Modeling and Monitoring of Pipelines and Networks*; Springer: New York, NY, USA, 2017.
4. Delgado-Aguiñaga, J.; Besancon, G.; Begovich, O.; Carvajal, J. Multi-Leak Diagnosis in Pipelines Based on Extended Kalman Filter. *Control Eng. Pract.* **2016**, *49*, 139–148. [CrossRef]
5. Santos-Ruiz, I.; Bermúdez, J.; López-Estrada, F.; Puig, V.; Torres, L.; Delgado-Aguiñaga, J. Online leak diagnosis in pipelines using an EKF-based and steady-state mixed approach. *Control Eng. Pract.* **2018**, *81*, 55–64. [CrossRef]
6. Sonnad, J.R.; Goudar, C.T. Turbulent flow friction factor calculation using a mathematically exact alternative to the Colebrook–White equation. *J. Hydraul. Eng.* **2006**, *132*, 863–867. [CrossRef]
7. Samadianfard, S. Gene expression programming analysis of implicit Colebrook–White equation in turbulent flow friction factor calculation. *J. Pet. Sci. Eng.* **2012**, *92*, 48–55. [CrossRef]
8. Brkić, D. New explicit correlations for turbulent flow friction factor. *Nucl. Eng. Des.* **2011**, *241*, 4055–4059. [CrossRef]
9. Winning, H.K.; Coole, T. Explicit friction factor accuracy and computational efficiency for turbulent flow in pipes. *Flow Turbul. Combust.* **2013**, *90*, 1–27. [CrossRef]
10. Zeghadnia, L.; Robert, J.L.; Achour, B. Explicit solutions for turbulent flow friction factor: A review, assessment and approaches classification. *Ain Shams Eng. J.* **2019**, *10*, 243–252. [CrossRef]
11. Azizi, N.; Homayoon, R.; Hojjati, M.R. Predicting the Colebrook–White friction factor in the pipe flow by new explicit correlations. *J. Fluids Eng.* **2019**, *141*. [CrossRef]
12. Pérez-Pupo, J.; Navarro-Ojeda, M.; Pérez-Guerrero, J.; Batista-Zaldívar, M. On the explicit expressions for the determination of the friction factor in turbulent regime. *Revista Mexicana de Ingeniería Química* **2020**, *19*, 313–334. [CrossRef]
13. Avci, A.; Karagoz, I. A new explicit friction factor formula for laminar, transition and turbulent flows in smooth and rough pipes. *Eur. J. Mech. B Fluids* **2019**, *78*, 182–187. [CrossRef]
14. Taler, D. Determining velocity and friction factor for turbulent flow in smooth tubes. *Int. J. Therm. Sci.* **2016**, *105*, 109–122. [CrossRef]
15. Popov, I.; Skrypnik, A.; Schelchkov, A. Generalized correlations for predicting heat transfer and friction factor of turbulent flow in tubes with inner helical ribs. *J. Phys. Conf. Ser.* **2019**, *1382*, 012032. [CrossRef]
16. Valve, C. *Flow of Fluids Through Valves, Fittings, and Pipe*; Vervante: Springville, UT, USA, 2013.
17. Menon, E.S. *Transmission Pipeline Calculations and Simulations Manual*; Gulf Professional Publishing: London, UK, 2015.

18. Gong, J.; Erkelens, M.; Lambert, M.F.; Forward, P. Experimental Study of Dynamic Effects of Iron Bacteria–Formed Biofilms on Pipeline Head Loss and Roughness. *J. Water Resour. Plan. Manag.* **2019**, *145*, 04019038. [CrossRef]
19. Wéber, R.; Hős, C. Efficient technique for pipe roughness calibration and sensor placement for water distribution systems. *J. Water Resour. Plan. Manag.* **2020**, *146*, 04019070. [CrossRef]
20. Santos-Ruiz, I. Colebrook Equation. Available online: https://zenodo.org/record/3348254#.X02dR4sRVPY (accessed on 31 August 2020).
21. Brown, G.O. The History of the Darcy–Weisbach equation for pipe flow resistance. In *Environmental and Water Resources History*; American Society of Civil Engineers: Reston, VA, USA, 2002; pp. 34–43.
22. Saldarriaga, J. *Hidráulica de Tuberías*, 3rd ed.; Alfaomega: Bogotá, CO, USA, 2016.
23. Weisbach, J. Lehrbuch der ingenieur-und maschinen. *Mechanik* **1845**, *1*, 434.
24. Marušić-Paloka, E.; Pažanin, I. Effects of boundary roughness and inertia on the fluid flow through a corrugated pipe and the formula for the Darcy–Weisbach friction coefficient. *Int. J. Eng. Sci.* **2020**, *152*, 103293. [CrossRef]
25. Swamee, P.K.; Jain, A.K. Explicit equations for pipe-flow problems. *J. Hydraul. Div.* **1976**, *102*, 657–664.
26. Haaland, S.E. Simple and explicit formulas for the friction factor in turbulent pipe flow. *J. Fluids Eng.* **1983**, *105*, 89–90. [CrossRef]
27. Serghides, T.K. Estimate friction factor accurately. *Chem. Eng.* **1984**, *91*, 63–64.
28. Lahiouel, Y.; Lahiouel, R. Evaluation of energy losses in pipes. *Am. J. Mech. Eng.* **2015**, *3*, 32–37.
29. Chin, D.A. *Fluid Mechanics for Engineers in SI Units*; Pearson: New York, NY, USA, 2017.
30. Levenberg, K. A method for the solution of certain non-linear problems in least squares. *Q. Appl. Math.* **1944**, *2*, 164–168. [CrossRef]
31. Marquardt, D. A method for the solution of certain problems in least squares. *J. Soc. Ind. Appl. Math.* **1963**, *11*, 431–441. [CrossRef]
32. Fletcher, R. A modified Marquardt Subroutine for Non-Linear Least Squares. Available online: https://epubs.stfc.ac.uk/manifestation/6683/AERE_R_6799.pdf (accessed on 31 August 2020).
33. Bermúdez, J.R.; López-Estrada, F.R.; Besançon, G.; Valencia-Palomo, G.; Torres, L.; Hernández, H.R. Modeling and Simulation of a Hydraulic Network for Leak Diagnosis. *Math. Comput. Appl.* **2018**, *23*, 70. [CrossRef]
34. Santos-Ruiz, I.; López-Estrada, F.R.; Puig, V. Estimation of Some Physical Properties of Water as a Function of Temperature. Available online: https://zenodo.org/record/2595273#.X02fLtozZPY (accessed on 31 August 2020).

© 2020 by the authors. Licensee MDPI, Basel, Switzerland. This article is an open access article distributed under the terms and conditions of the Creative Commons Attribution (CC BY) license (http://creativecommons.org/licenses/by/4.0/).

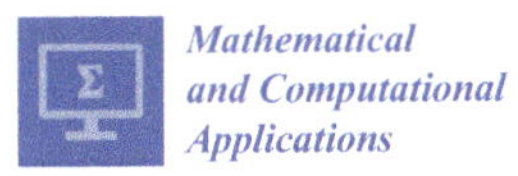

Article

Sensorless Speed Tracking of a Brushless DC Motor Using a Neural Network

Oscar-David Ramírez-Cárdenas [1] **and Felipe Trujillo-Romero** [2,*]

[1] División de Estudios de Posgrado, Universidad Tecnológica de la Mixteca, Carretera a Acatlima Km 2.5, Huajuapan de León 69000, Oaxaca, Mexico; odramirez@mixteco.utm.mx

[2] División de Ingenierías, Campus Irapuato-Salamanca, Universidad de Guanajuato, Carr. Salamanca-Valle km 3.5+1.8, Comunidad de Palo Blanco, Salamanca 36700, Guanajuato, Mexico

* Correspondence: fdj.trujillo@ugto.mx; Tel.: +52-464-647-9940

Received: 30 July 2020; Accepted: 2 September 2020; Published: 4 September 2020

Abstract: In this work, the sensorless speed control of a brushless direct current motor utilizing a neural network is presented. This control is done using a two-layer neural network that uses the backpropagation algorithm for training. The values provided by a Proportional, Integral, and Derivative (PID) control to this type of motor are used to train the network. From this PID control, the velocity values and their corresponding signal control (u) are recovered for different values of load pairs. Five different values of load pairs were used to consider the entire working range of the motor to be controlled. After carrying out the training, it was observed that the proposed network could hold constant load pairs, as well as variables. Several tests were carried out at the simulation level, which showed that control based on neural networks is robust. Finally, it is worth mentioning that this control strategy can be realized without the need for a speed sensor.

Keywords: PID control and variants; Intelligent control techniques; neural control; brushless DC electric motors; sensors and virtual instruments; analysis and treatment of signals

1. Introduction

Nowadays, Brushless direct current (BLDC) Motors are extensively used because of their characteristics. Such characteristics are high dynamic response and high-power density. Usually, these kinds of motors are controlled employing an electronic inverter of commutation composed of three-phases. Typically for controlling these types of motors, a six-step commutation and a three-phase voltage source inverter are used, where the commutation of the inverter components is determined by the state of the Hall effect sensors. In recent years, works have been reported for speed control in this type of engine based on this scheme; such is the case of the work presented by Zhao et al. [1].

There have been considerable efforts in solving the soft computing approach to control brushless motors. The work developed by Shanmugasundram et al. [2] presents a comparison of control for a brushless motor using PID, fuzzy logic, and a hybrid of neural networks which show the effectiveness of these controllers such as speed response, steady-state error, among others. Arulmozhiyal and Kandiban [3] made a fuzzy PID controller in which they present the facility to tune the PID constants using the implemented fuzzy system. For their part, Premkumar and Manikandan developed a diffuse neuro controller, which had a higher performance than the other implementations tested [4]. In [5], Al-Maliki and Iqbal have also used a fuzzy Logic Control, but this time for tuning a PID controller. In addition, Mamadapur and Mahadev [6] implemented a PID for speed control of a BLDC motor but using Neural networks.

For their part, Liu et al. [7] developed a control strategy for these types of motors that consisted of the implementation of control through an adaptive neural network for current monitoring in this system. Ibrahim et al. [8] made a comparison between the Particle Swarm Optimization techniques and the bacterial food search technique to determine the optimal parameters of a PID speed controller for a BLDC motor. In [9], an adaptive fuzzy logic for speed control of a BLDC motor was developed using ANN and PSO techniques. Potnuru et al. [10] used a pollination algorithm for speed control of a BLDC motor. Another work that implements a speed control for a BLDC motor can be found in [11], where an adaptive neural network with a fuzzy inference core was used. Templos-Santos et al. have used bio-inspired algorithms in order for parameter tuning of a PI controller in a Permanent magnet synchronous motor (PMSM) [12]. Merugumalla and Kumar [13] used the firefly algorithm to create a motor drive of speed control for a BLDC. One last work is that developed by ELkholy and El-Hay [14], and they have evaluated the efficient dynamic performance of BLDC motor using different soft computing approaches. It is worth mentioning that the works mentioned above use a sensing mechanism to give feedback to the control system, making this more elaborate and, in some cases, more costly to implement these methods.

More focused works concerning the control of direct current motors, beginning with the work of Yu et al. [15], who used a neural network to carry out adaptive control in a PMSM. For their part, Cheng et al. [16] made use of a radial neural network for the control of a brushless DC motor with excellent results. The fusion of neural networks with a diffuse system was presented by Abed et al. [17] to diagnose failures in the bearings of a BLDC engine. In [18], Luo et al. performed the control of a brushless motor using the adaptive dynamic surface control technique modeled by a neural network. Saleh et al. [19] used wavelet neural networks to implement a speed controller for a BLDC motor. Naung et al. [20] adjusted the parameters of a controller PI to control the speed of a BLDC motor through a neural network. In [21], Dynamic Neural Networks were used to develop an intelligent control for a BLDC Motor. Finally, let us mention the work of Ho et al., who implemented a driver for a DC motor through the use of neural networks [22].

In this paper, the main motivation to study the sensorless control of a BLDC motor is the use of neural networks.The above is due to the neural networks are a better way to recognize patterns from the sample of the system which going to be controlled. Therefore, when the control of a BLDC motor wants to be made, usually, the use of observators is necessary in order to estimate, mainly, the position of the motor. These observators replace the use of sensors but are very complicated to implement. For this reason, the use of any sensorless techniques for control or estimation of the parameters of the motor. For example, in [23], the five-phase induction of the motor was modeled by means of the short-circuit fault between turns and sensorless control strategy. In [24], König, Nienhaus, and Grassohave analyzed and modeled the current ripples generated by electromagnetic actuators to estimate the inductance and carry out sensorless monitoring of the device. On the other hand, Che et al. [25] implemented a sensorless speed control for an induction motor using the sliding modes method and genetic algorithms. Another kind of motor where the sensorless technique was used is the Permanent magnet synchronous motor. In the work of Wu and Zhang [26], sensorless speed control was developed for a PMSM using terminal sliding mode and backstepping. In [27], Aguilar et al. implemented a sensorless speed tracking controller for a PMSM based on a second-order sliding mode observer and tested with load variations. Meanwhile, Kivanc and Ozturk have developed a position sensorless speed control for PMSM by means of a space vector based on four switches and three phases inverter [28]. In [29,30], interesting studies about sensorless control of PMSM are presented using estimation methods and sliding mode observer.

Next, a couple of works where the sensorless and neural networks are combined will be mentioned. One of them is [31], where a sensorless speed system was implemented based on the reactive power-based model reference adaptive system speed estimator and adaptive neural network for a PMSM. In [32], a neural network was used for a sensorless control for a PMSM oriented tor wind energy conversion.

On the side of the BLDC motor, Sreeram [33] implemented a speed regulation control of a sensorless BLDC utilizing fuzzy logic and four-switch three-phase inverter. The development presented in [34], where the radial displacement of the stator is proposed as a mechanism to improve speed control for a BLDC drive in a sensorless way. In [35], Vanchinathan and Valluvan used the Bat algorithm to control in a sensorless way the rotor speed of the BLDC motor. A study comparative of different controller techniques is presented in [36], where a controller PI, an anti-windup PI, a fuzzy logic-based, and a fuzzy controller PI are compared to validate the performance of sensorless BLDC motor at different loads and speeds. Rif'an, Yusivar, and Kusumoputrohave proposed the use of the ensemble Kalman filter (EnKF) and neural networks to predict load changes and estimates of the disturbance by simulation of a sensorless BLDC [37].

Finally, other works that made use of non-invasive methods for sensing some physical variable or detect failures in motors will be next mentioned. In [38], the technique of Signal Analysis based on Chaos using Density of Maxima(SAC-DM) technique was applied in the diagnosis of failures of BLDC motors from sound signals. In addition, Medeiros et al. [39] used the SAC-DM to detect failures in BLDC motors. Meanwhile, other works are focused on detecting faults in induction motors based on the analysis of acoustic sound and vibration signals [40–42].

For this reason, in this work, the controlling brushless DC motor through neural networks and without the need for sensing is proposed. Solving this kind of problem is essential since it is not always easy to have a sensor for the different internal parameters of a motor in order to control such parameters, especially for a BLDC driver. In addition, the neural network utilized to solve the controller of this motor is implemented only with two-layer; then, the neural network is simple and easy to port into embedded hardware to generate a non-dependent computer system.

In order to show a generalized idea of the advantages of our implementation, the highlights of our work will be listed below:

(i) A sensorless velocity control was developed through a neural network of the type multi-layer perceptron architecture.
(ii) The neural network implemented has only two layers. This means that the network is the smallest of such used in the majority of the papers.
(iii) The training set used is formed of just 22 values being that there are 5000 example data extracted from the curves used for training.
(iv) The use of a neural network applied to the control of a motor avoids the use of a sensor.

The main novelty of our approach is found in the third item of the advantages previously presented, due to the vast majority of works about this topic and that use a neural network utilizing a big number of examples to train the neural network.

The implementation and the results obtained in various tests carried out are presented in the following sections—starting with Section 2, where the typical structure for brushless motor control is shown as well as its simulation in the PSIM software [43]. In addition, in Section 2, a brief review of the neural network and its architecture used is presented. The implementation of motor control and the training of the neural network are described in Section 3. Section 4 shows the tests and results of the motor control implementation through the neural network. Additionally, a comparison against the PID control is made in Section 4, too. In order to clarify some points about the results obtained in this work a discussion was added in Section 5. Finally, the conclusions of this work, as well as perspectives for its continuation, are presented in Section 6.

2. Materials and Methods

This section presents and explains the different methods and materials used to achieve the results in this paper. First, it is necessary to present a brief description of the brushless direct current just to contextualize the functioning of the motor. In the second part of this section, a general revision about the concepts of neural networks will be presented.

2.1. Motor Brushless

A brushless motor is an electric motor controlled by an electrical signal. This type of motor lacks any form of collector or sliding ring. The motor requires some form of alternating current to rotate, either from an alternating current (AC) power supply or an electronic circuit. These motors have certain advantages over brushed motors, which are ideal for robotics projects. Some advantages and disadvantages of this type of motors are presented below:
Advantages:

(a) Higher efficiency (less heat loss)
(b) Higher performance (longer battery life for the same power)
(c) Less weight for the same power
(d) They require less maintenance since they do not have brushes
(e) Speed/torque ratio is almost a constant
(f) Higher power for the same size
(g) Better heat dissipation
(h) High speed range without mechanical limitation.
(i) Less electronic noise

Disadvantages:

(a) Higher construction cost
(b) Control is through an expensive and sophisticated circuit
(c) It always takes an electronic control to work (ESC's), which sometimes doubles the cost.

The BLDC motor uses a DC power supply applied by power devices, and the position of the rotor determines its switching sequence. The BLDC motor phase current, which has a rectangular waveform, must be synchronized with the rear counter electromotive force (EMF) to produce a constant torque at a constant speed. Electronic power switches replace the conventional mechanical switch, and the current of the motor windings is supplied depending on the position of the rotor. This type of AC motor is called a brushless DC motor, and its operation is similar to that of conventional DC motors with a switch.

BLDC motors are generally controlled employing three-phase inverters, so it requires a rotor position sensor to control the inverter in order to determine the switching instants and provide an appropriate sequence for switching. The position of the motor in these engines is generally determined by position sensors such as Hall effect sensors, resolvers, and absolute position sensors; otherwise, it must be prepared by estimating the parameters that specify the rotor position using sensorless techniques [44,45].

The driving range for each phase is usually 120 electrical degrees, which is divided into two 60-degree steps. For example, S5-S6, S1-S6, S1-S2, S3-S2, S3-S4, and S5-S4 are six steps of a 360-degree electric cycle. Therefore, only two phases are carried out at a time. The above can be seen graphically in Figure 1.

From the operation of the engine and their elements, the BLDC engine is modeled to simulate its behavior. The brushless motor model implemented in the PSIM simulation environment is shown in Figure 2. This figure shows the three-phase inverter, which has a sensor to monitor the current supplied to the BLDC. It is also possible to see that the BLDC motor is coupled with a sensor for speed reading. This sensor generates the control signal that determines the pulse width of the PWM (Pulse Width Modulation) that activates each of the inverter's drivers, in addition to the necessary switching stage that depends on the state of the Hall effect sensors. This switching allows the three-phase inverter to function normally in addition to preventing a short circuit in any of the three branches.

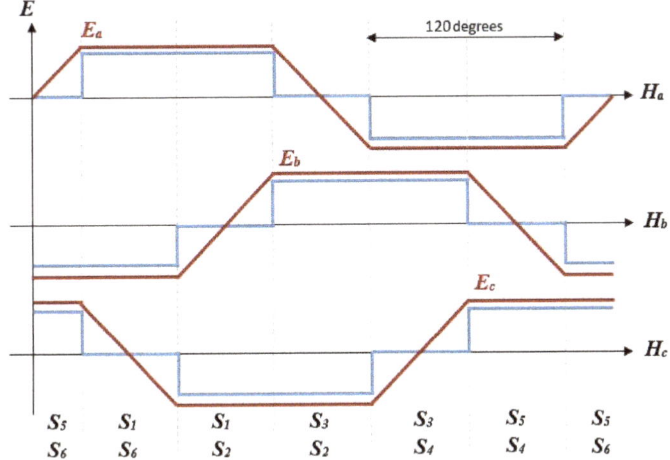

Figure 1. Waveforms of the counter-electric force in 120 electrical degrees.

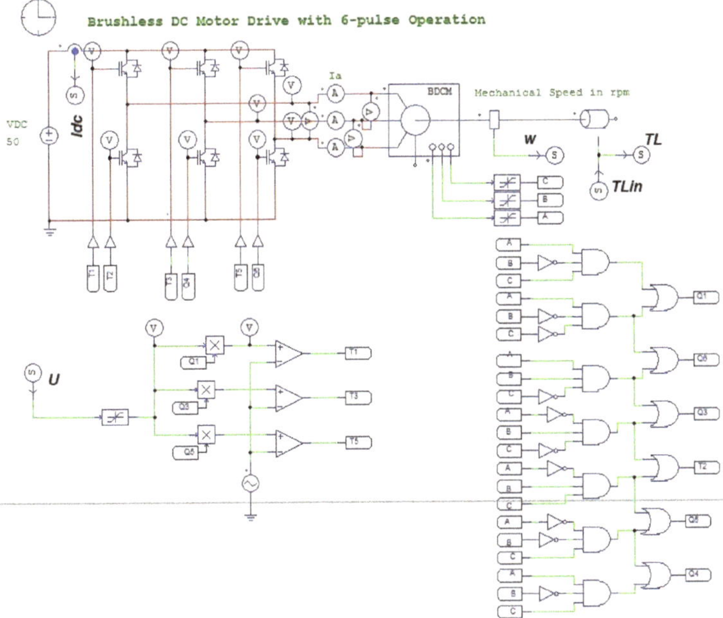

Figure 2. Inverter-BLDC system implemented in PSIM.

2.2. Neuronal Networks

For artificial intelligence, neural networks are a significant field of application. These are inspired by the interconnections of the cells of the human brain. In addition, they try to simulate the behavior of a biological neural network. Thus, neural networks are usually used to model systems that solve difficult problems through other techniques.

Some examples where artificial neural networks have been used to solve various tasks such as pattern recognition [46], parameter estimation [47], prediction of values [48] and of course in the control of different types of systems [49–51].

In general, a neural network is formed by the next elements:

1. Neurons
2. Layer
3. Transfer functions
4. Weights

Usually, the neural networks have several layers composed of a different number of neurons each of them. In general, neural networks are composed of three different layers: input, hidden, and output. The information goes from one layer to another through the transfer functions. The transfer function determines how the information is transmitted between layers. Weights are directly associated with the connection between layers. That is, if a connection has weight with a high value, then it will make more contributions to the network output.

One of the most used neural networks is called multi-layer perceptron, and the best-known training method for this type of network is backpropagation. This algorithm is used more for learning this kind of multi-layer perceptron network. The backpropagation algorithm trains the neural network from sample vectors of the system that are of interest for their modeling, such as the texture of an object or, in our case, the speed of a motor.

Summarizing, the backpropagation training system consists of the following steps:

1. Initialize the weights of the network randomly.
2. Enter input data from among those to be used for training.
3. Let the network generate an output data vector (forward propagation).
4. Compare the network output with the desired output.
5. The difference between the generated and the desired output (called error) is used to adjust the weights in the output layer.
6. The error spreads backwards (hence the name of backpropagation), towards the previous neuron layer, and is used to adjust the weights of that layer.
7. Continue propagating the error backwards and adjusting the weights until the input layer is reached.

One of the advantages that backpropagation presents is a speed of convergence and robustness compared to other types of training. For this reason, in this work, a multi-layer perceptron neuronal network has been implemented using backpropagation as a training algorithm. This network was used for speed control of a brushless motor. The architecture of the neural network used for the development of this work has a topology 2:8:1 and uses only two layers. The relation 2:8:1 means that the neural network has an input vector of a size of two elements. This means that every example presented to the neural network only has two elements. The second number 8 belongs to the number of neurons that have the first hidden layer. Finally, number 1 tells us that there is only one neuron in the output layer.

As it is possible to see, the neural network needs only two layers of neurons in order to obtain an excellent performance of the control system proposed. However, just with this configuration, the neural network is capable of learning the vectors related to brushless motor control.

Part of the neural network training will be addressed in the next section.

3. Neural Network Training

As previously noted, it is necessary to have example vectors related to the problem to be solved by the neural network. These sample vectors must come from real examples of the objects or the system to be modeled. In this case, examples of the motor's response to different applied torques are required.

For this reason, a classical Proportional, Integral, and Derivative (PID) controller has been implemented in order to obtain previous test values to train the proposed network. Figure 3 shows the block diagram of the PID controller used. This figure consists of a DC/AC converter, usually called a voltage inverter. This inverter supplies the three-phase voltages for the BLDC motor. The applied load torque, the sensing of the rotor speed, a reference speed block that will serve for the training of the network, the PID control, and the inverter switching stage determined by the states of the Hall effect sensors are detailed in Figure 2.

From the PID control shown in the diagram in Figure 3, the different values for the input vectors of our neural network are obtained.

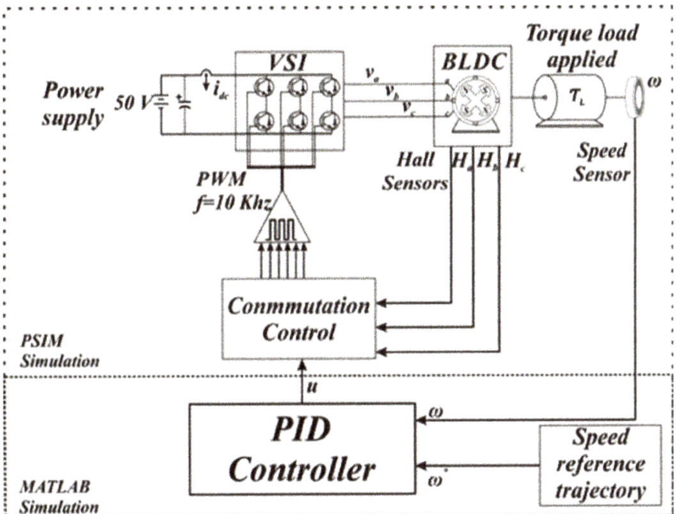

Figure 3. Block diagram of the PID control used.

Figure 4 shows the training signal made to the BLDC motor with the PID control. This signal must follow the desired reference that starts from scratch and increases its speed smoothly to 750 rpm (revolutions per minute). It is worth mentioning that the value of 750 rpm is the maximum speed that the motor can reach. In the same figure (Figure 4), it is possible to observe the signal resultant, which was obtained with no external load pair applied to the system. For this reason, it will be mentioned as a zero pair in future mentions of it. In order to have greater control over the gains of the PID controller, they were chosen heuristically. Hence, the chosen earnings were 5, 0.001, and 0.0001 for Kp, Ki, and Kd, respectively. From this test, pairs of velocity values (ω) and the control signal (u) necessary to reach the velocity values for each pair (ω, u) are taken. These values are sampled every 40 rpm; therefore, 600 input vectors will be used as examples for the neural network. These values will serve to train our neural network so that it is capable of controlling the motor with zero torque.

Figure 5 graphically shows the 22 values obtained from the zero torque PID control tests that will be used to train the neural network. It is worth mentioning that, within the sample pairs used, the minimum and maximum values are included, which are 0 and 800 rpm, respectively.

Obtaining the previous values, as can be seen in the previous paragraphs, is only to train the neural network when the motor does not have an applied load, that is, when the torque is zero. Therefore, to determine the behavior of the network when a non-zero torque is applied, it is necessary to train the network with different load values.

As it was required to have the performance of the neural network in a range of 0 to 1 Nm, it was considered to train the network with load pairs applied in the said range in increments of 0.25 Nm.

To obtain the input vectors for each of the different load pairs to be learned by the network, the same procedure is performed as for obtaining the vectors with zero torque. Obviously, this time the network is trained with the desired speed values, control signal, and applied torque (ω, u, τ) for each applied torque from 0 to 1 Nm in increments of 0.25 Nm.

Some values obtained for each of the different applied load pairs (τ) are observed in Figure 6. These values are used to train the network that can control the BLDC system when any load pair is applied.

This will allow the neural network to have a dynamic performance when applying a load in the range of 0 to 1 Nm. Although the network will train only with well-defined load values every 0.25 Nm, it will be able to infer the performance that it should have in the desired speed values that were never presented to the system.

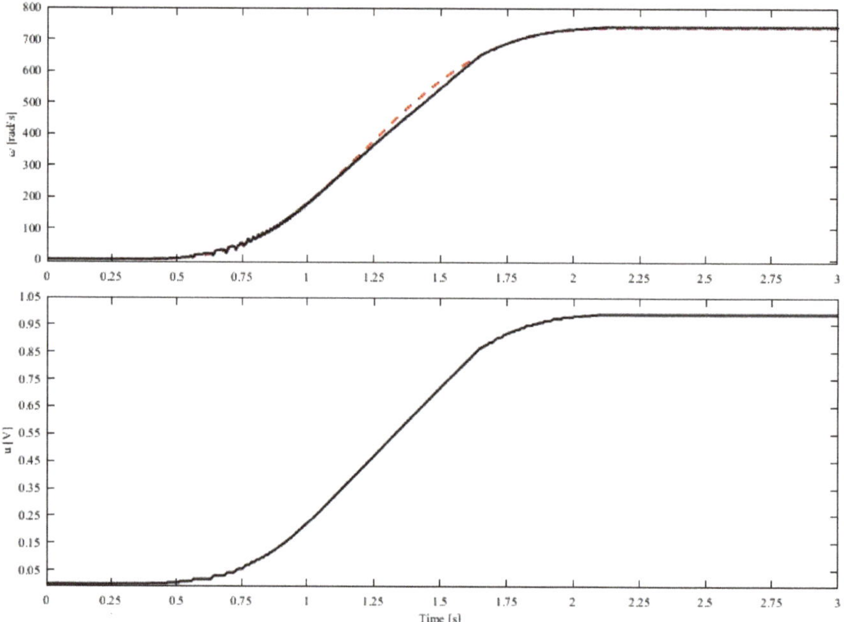

Figure 4. Training signal without torque load applied.

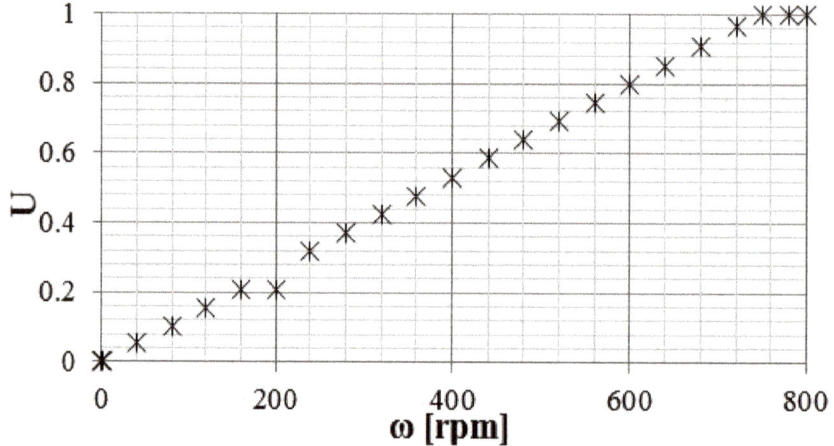

Figure 5. Values obtained from the PID controller ($\tau = 0$).

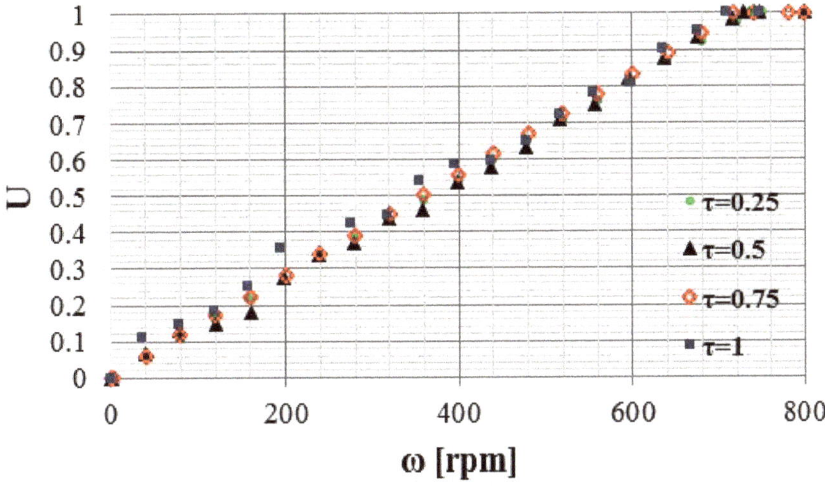

Figure 6. Training values obtained from PID control with different loads applied.

Like any control system, it is necessary to determine the parameters with which the controller will work. Unlike a PID controller, a neural network does not require that these values be modified every time any element of the system changes. This is another advantage of using a neural network compared to classical control methods. This is because the network adapts to resolve cases that were not presented during training. It should be mentioned that this only happens within the logical ranges of the expected behavior of the system.

In this case, the parameters of the neural network and its training are presented in Table 1.

Some observations regarding the values shown in Table 1. First, it will be mentioned that, although the neural network training indeed used the backpropagation algorithm, this algorithm has four variants. These variants are listed below:

1. Momentum,
2. Variable learning ratio,
3. Conjugate gradient, and
4. Levenberg–Marquardt.

Table 1. Neural network parameters.

Parameter	Value
Number of layers	2
Input layer neurons	8
Activation function	Sigmoidal
Output layer neurons	1
Activation function	Lineal
ANN architecture	Multi-layer perceptron
Training algorithm	Backpropgation
Variant	Levenberg–Marquardt
Performance	Mean Squared Error
Epoch	5000

Being the Levenberg–Marquardt algorithm, the backpropagation variant with which, generally, better performance is obtained in the tasks assigned to the neural network. For this reason, in this implementation, it was decided to use this variant. Although this variant indeed has the highest computational cost, it is compensated by the excellent performance it offers.

Another observation regards the number of epochs that were specified as the maximum value in this implementation. In general, for a neural network to converge towards the optimal values of the weights, 1000 to 10,000 epochs are required. For this reason, this parameter began to carry out tests from 5000 epochs. However, in the different experiments made, it was observed that the neural network had a convergence in a range of 18 to 50 epochs. The last-mentioned means that training does not take a long time of 50 epoch to obtain a result. Then, this parameter was established in 50 as a maximum value that training needed to reach the optimal values of the weights.

The above is due in part to three main things. One of them is the use of the Levenberg–Marquardt algorithm for training, which has a faster convergence towards the optimal values of the network weights even if it requires more processing to get convergence of the network. The other is the number of network layers, which, in our case, only have the input and output layers.

Finally, the number of neurons is mentioned; in our implementation, only nine neurons are necessary. Eight of these neurons are in the input layer, while the output layer only has one neuron.

It is important to emphasize that our implementation occupies few neurons, as previously mentioned. This is a feature that will allow us to implement this controller quickly and efficiently in an embedded system, such as an FPGA [52]. It would even be possible to implement it on a lower-performance device such as an Arduino. The above would give portability to this controller, thus obtaining a lower-cost hardware implementation.

Once the implementation of the brushless motor controller through neural networks is presented, the results obtained are shown. Therefore, in the following section, both the tests performed and the results obtained from this implementation will be presented.

4. Results

This section presents the different tests carried out on the implementation of the control performed with the neural network as well as the comparison against the control carried out by the PID. It will begin with the performance of the control performed by the neural network when there is no load applied to the plant.

The different tests and results were developed using the Matlab/Simulink-PSIM package. For this, the BLDC model SG/F14 Gearless motor parameters were used. Parameters mentioned were obtained by characterizing it and used in the PSIM simulation environment. These values are shown in Table 2.

Table 2. Motor parameters.

Parameter	Value
Stator resistance (R_s)	0.15 Ω
Stator self-inductance (L_s)	918×10^{-6} H
Stator mutual induction (M)	-367.2×10^{-6} H
Vpk/krpm	78.9 V
Vrms/krpm	55.86 V
Number of poles (P)	30
Moment of inertia (J)	1.5855×10^{-3} kgm^2
Mechanical time constant (τ_{mech})	0.2369

Figure 7 shows the block diagram of the plant which is made up of the following elements:

1. A three-phase BLDC motor
2. A mechanical load
3. A Voltage Source Inverter (VSI)
4. A current sensor (to measure the direct current (i_{dc}) consumed by the inverter)
5. A PWM device (with a frequency of 10 kHz)

6. The path speed reference
7. The neural network implemented in Matlab/Simulink.

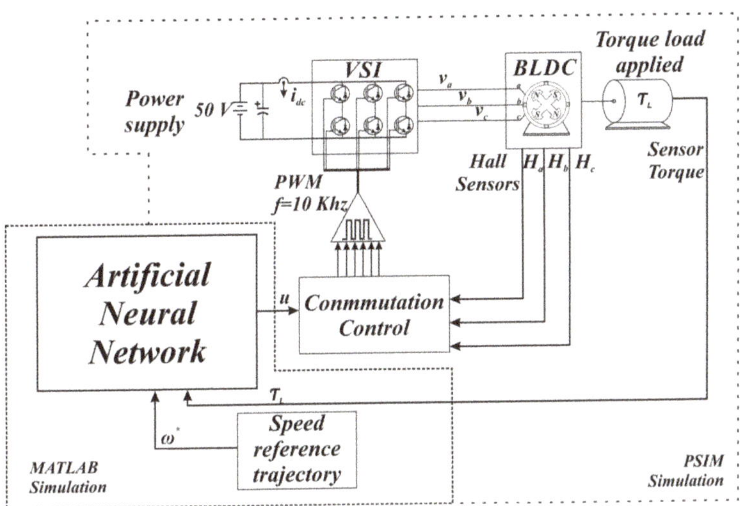

Figure 7. Configuration diagram for the BLDC system.

On the other hand, Figure 8 shows the diagram of the implementation of the neural network made in Matlab/Simulink. In this diagram, the stage that defines the desired path, the applied load torque, as well as the implemented neural network and the BLDC motor system can be seen, shown above in Figure 2.

It will start with the performance of the control performed by the neural network when there is no load applied to the plant. The behavior of the trained network can be seen in Figure 9. In this test, the neural network is subjected to following a desired speed reference path from zero to 650 rpm. It should be mentioned that this was not a network training value. However, the fact of training without a load applied to the plant was to have a reference value for later comparison with variable load.

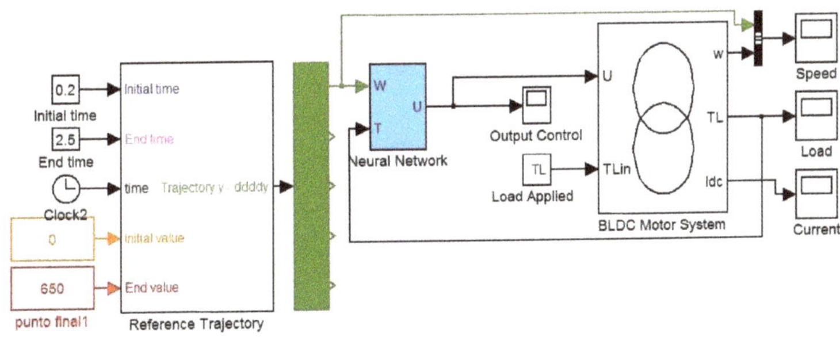

Figure 8. Implementation of the Matlab/Simulink-PSIM simulation system.

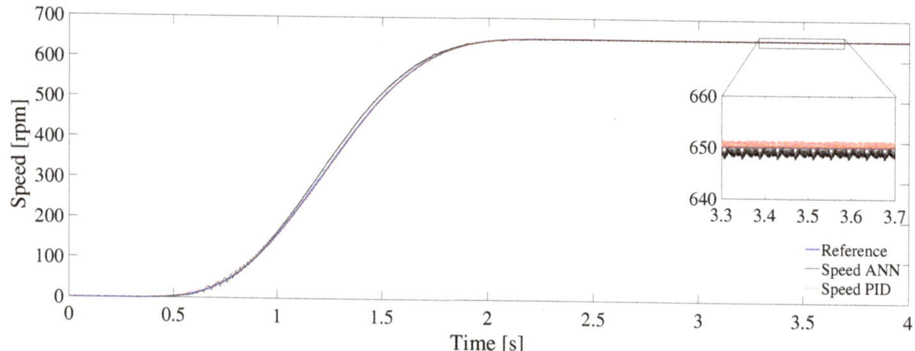

Figure 9. Speed response of the neural network and the PID with no load applied to the plant.

Furthermore, in this same figure, it can be observed the comparison of the control signals of both the network and the PID control. This comparison shows that both velocity signals are very similar, but the neural network remains around the desired reference. With this, it is shown that the scheme of the proposed network is functional.

The behavior of the implemented neural network control signals compared to that of the PID is shown in Figure 10. In this figure, it can be seen that the two responses to the desired velocity values are very similar.

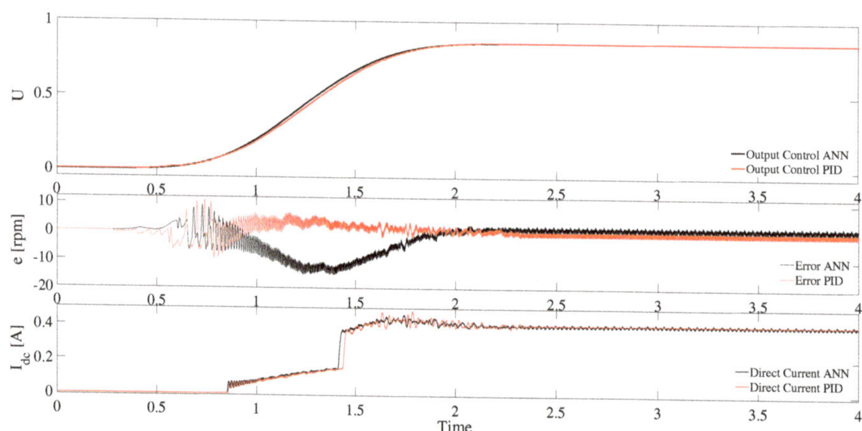

Figure 10. Control signals, speed, and current error on the controller CD bus with the neural network and with the PID.

In the same Figure 10, a satisfactory behavior of the neural network is observed both in the speed error obtained and the current required by the inverter. Analyzing the system error and leaving aside the part of the transient. In this case, it can be seen that the neural network has a much smaller error than that obtained by the controller through the PID; furthermore, the neural network is maintained around the desired reference.

On the side of the current required by the inverse, it can be seen that the one required with the control of the neural network is more stable and constant, unlike the requirement of the PID.

With the finality of extending these results, a test was also performed with a constant load torque applied to the BLDC motor and variable load torques.

Having seen the performance of the systems under conditions known, as are the values close to those of neural network training, it was proceeded to evaluate the system control with values different from these. In this test, the system was subjected to a load torque. The applied load was 0.3 Nm at $t = 2.5$ s.

Under these characteristics, Figure 11 shows the speed monitoring from a speed of 0 rpm to 650 rpm. In this figure, it is possible to emphasize that the neural network is capable of compensating the speed response to this applied load torque. It should be noted that this applied torque was not used for network training.

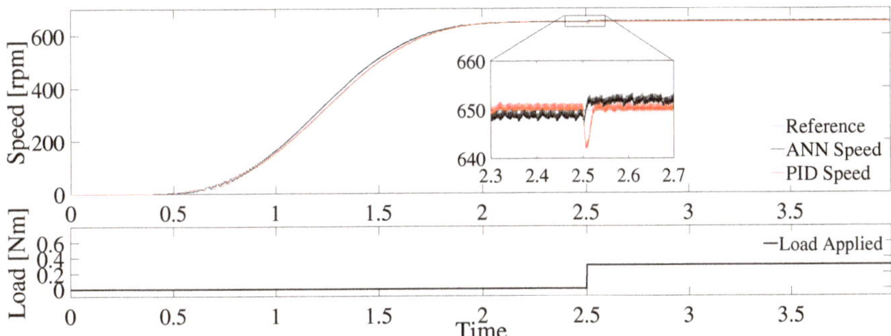

Figure 11. Speed response of the neural network and the PID to two applied load changes.

Although the neural network indeed maintains an error with respect to the reference value, especially after the application of the load, it can be seen that the response of the network is smoother and without transients that could affect the performance of the plant as with the PID response.

Continuing with the analysis of this test, Figure 12 shows the control output of the neural network, which minimizes the speed error. Here, we also observe the comparison of the current consumed by the inverter for both the grid and the PID.

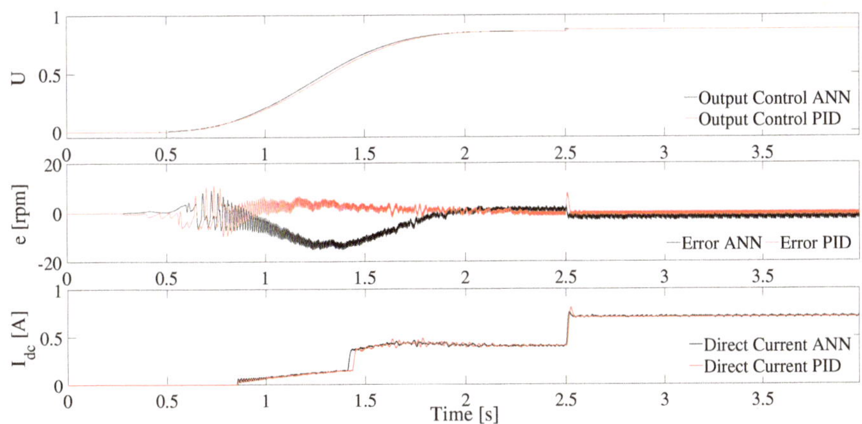

Figure 12. Control signals, speed, and current error on the CD bus for the neural network and the PID.

An important point that should be mentioned is that, in this implementation of the control by the neural network, the output speed of the system is not given as feedback. Therefore, this neural network control can be considered as a sensorless type system. The latter is due to it not being necessary to measure the speed for the system to work acceptably, with the inclusion of a system that does not require excessive computational load, and that does not use a sensor to measure speed. This allows us to make the implementation less expensive, mainly in computational terms. Therefore, the use of a neural network for the control of this type of motors is a highly recommended element to carry out the control.

Finally, a test was performed where the system was subjected to the same desired speed path as the previous tests. However, to carry out this test, the system was subjected to a variable load torque. This torque load varied from 0 to 0.7 Nm in an irregular way and can be seen in the lower part of Figure 13.

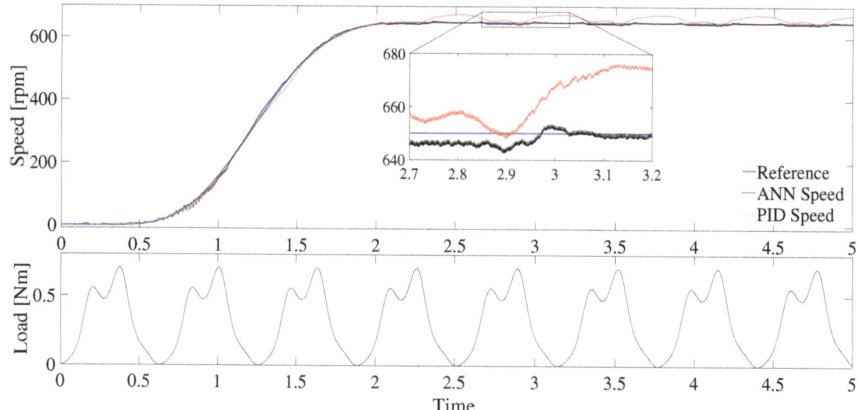

Figure 13. Speed response of the neural network and the PID by applying a variable load over time.

This experiment gave us an interesting result and is the one presented in Figure 13. In this figure, it is observed that, despite the variable load torque, the system implemented with the neural network is capable of responding to this load variation. However, the PID controller fails to respond to these load variations adequately.

The reason the PID is unable to respond to changes in applied load variation is that the PID control signal becomes saturated and prevents it from reacting adequately to variable torque changes.

Another important point to highlight in this test is that the error generated by the neural network is less and closer to the value of zero than that obtained by the PID. This can be seen in Figure 14.

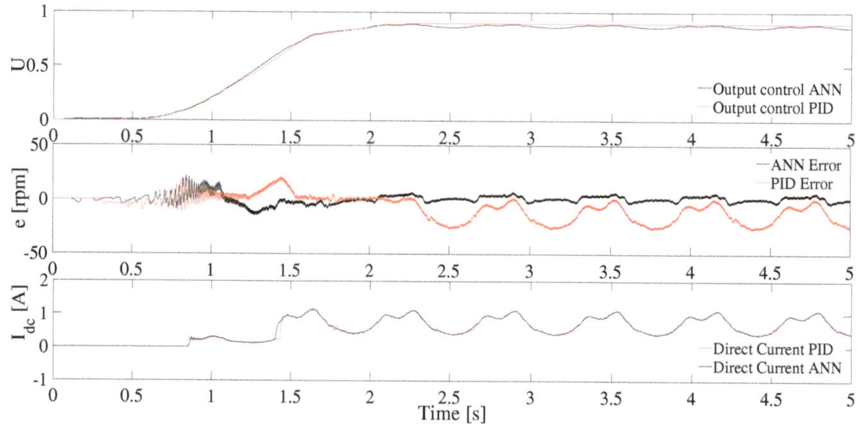

Figure 14. Speed response of the neural network and the PID by applying a variable load over time.

Controller Performance

Sometimes, it could be challenging to follow the graphics where the results are shown. The graphics are indeed a better way to present information but tiny, and essential details, are possibly not highlighted.

For this reason, in this section, some specific parameters like the controller performance, the power consumption, and the mean square error will be presented to give another point of view of the advantages in the approach presented in this work.

Some notes about the tables are next. The test marked as "No-load applied" belongs to the experiment shown in Figures 9 and 10, while the legend "Constant load applied" is for the results in Figures 11 and 12. The results presented in Figures 13 and 14 are labeled as "Variable load applied."

Beginning with our explanation, Table 3 shows the performance developed by each controller. Here, it is possible to see the different tests done and the performance obtained in every case. In this Table 3, it is easy to observe that the performance carried out by the two controllers is so close. However, the controller implemented through neural networks is robust since its performance, in the three cases, is around 99%. Meanwhile that made by the PID begins to show a dispersion when the load is applied to the motor varies.

Table 3. Controller performance.

LOAD	PID	ANN
No-load applied	0.1186%	0.0332 %
Constant load applied	0.1351 %	0.0930 %
Variable load applied	2.9911 %	0.3147 %

Concerning the mean square error, similar behavior is observed. In this case, Table 4 shows the way that the error grows for the two controllers as the load applied to the BLDC motor varies. From the results presented in Table 4, it is evident that the values shown are high. This is due to the oscillations generated by the controllers when trying to follow the desired speed. Even so, the fact that the controller made by the neural network has a tendency to be more close to the desired value must be highlighted since the mean square error values in all three cases are lower than in the PID controller.

Table 4. Mean square error for the controller implemented.

LOAD	PID	ANN
No-load applied	952.18	425.4980
Constant load applied	1132.7	514.3944
Variable load applied	2794.2	946.8229

Finally, Table 5 exhibits the values corresponding to the average power consumption for each controller in the different tests carried out. As a reference value, the BDLC motor used for the experiments in this work can deliver a maximal power of 50 Watts. Thus, the values that are shown in Table 5 tell us that the PID controller wears more power than the one made using the neural network. In the test with a no-load applied, the power consumption in both controllers has almost the same value for both PID and Neural network controllers with just a minimal variation of 0.0529 Watts. In the other two tests, the power that is expended by the PID controller begins to grow; meanwhile, the power expended by the neural network maintains stable growth without significant variations. Here, again, the neural network controller is better than the PID controller.

Table 5. Average power consumption

LOAD	PID	ANN
No-load applied	14.4753 W	14.4224 W
Constant load applied	27.127 W	21.72 W
Variable load applied	32.2768 W	25.8408 W

5. Discussion

In order to clarify some points about the results obtained, a discussion will be presented in this section.

One of the considerations that it is necessary to comment on is the fact that, when a BLCD motor does not have Hall sensor, the counter electromotive force detection methods are commonly used to know the position of the motor at all times.

However, when the BLCD is fixed, no counter electromotive force is generated; therefore, there is no information on the position of the stator and rotor. The above mentioned is why motor starting methods are used, such as starting the motor in an open-loop configuration by activating the coils in a predetermined sequence. Now, when using Hall sensors, the use of digital tachometers based on the time of the rising edge of one of the Hall sensors or use of external sensors dedicated to speed reading, such as encoders or resolvers, solves the problem of starting and running the engine at low speeds.

Therefore, this work addresses the use of Hall sensors to control switching by varying the PWM output signals that go to the driver (see Figures 3 and 7).

Another consideration is that the term sensorless refers to the fact that, in this approach, no external sensor or digital tachometers are used; that is, the proposed system only makes use of the desired speed and applied torque.

The next point to take into account is the expected performance of the control implemented using neural networks. It is evident that there will almost always be differences between real models and those used experimentally on simulations.

Independently of the method used to simulate a system like that shown in this work, it is almost certainly necessary to make adjustments in the real implementation. For this reason, the logical step to develop is to retrain the neural network. However, some points must be taken into account. First, the fact of the vectors for training the neural network to develop the proposed control were obtained from the real parameters of the BLDC motor target in this work. The above gives us an idea of the behavior of the real from the simulation tests. The physical parameters and variation due to wear are indeed an essential variable in the real motor. This is going to have a direct impact on the BLDC motor performance. However, the behavior of the real is expected to be closely similar to simulation. From the experience of when other control approaches have been applied, similar results have been obtained both in simulation and in the real system, especially in terms of the speed and current values obtained.

However, the behavior of the real and simulation tests is expected to be very similar since different control approaches have been applied, obtaining similar simulation and real results in terms of speed and current.

Finally, one last question arises: what if the BLDC motor changes? Here, the expected response of the neural network will be close to the training done. This means that the neural network will be able to always control the motor, and the parameters of the BLDC motor are near the value used to train in our work. The above mentioned is due to the ability of the neural networks to generalize information from data learning. It is an advantage of the neural networks facing other methods like PID, since the neural network will only need to be trained if the parameters of the motor to be controlled are very different. While a PID is just one variable changes slightly, the system needs to be tuned again.

6. Conclusions

In this work, an implementation of the speed control of a BLDC motor using a multi-layer perceptron type neural network was presented. From the tests carried out in this article, it was observed that the control implemented from the proposed neural network scheme is efficient, and satisfactory results similar to those presented by a PID in the presence of constant load pairs are obtained.

According to the results obtained in the different tests and experiments carried out in our implementation, it was evident that the neural network is superior to a PID control. This is because, in most real applications, it is difficult to find a constant load torque; therefore, the use of a PID control may not be appropriate for a system. This is highlighted by the fact that the PID control is not capable of supporting variable load changes. At this point, the neural network was superior since it is capable of supporting torque changes without the need for a speed sensor.

Finally, it is noted that the implementation of the neural network is computationally less expensive since it only needs nine neurons distributed in two layers. This will allow the future implementation of this control in embedded hardware.

Among the perspectives considered for the continuation of this work, just two of them will be mentioned—first, to implement this controller in an embedded system to validate its performance in a real system. Secondly, the improvement-oriented towards obtaining a system that does not depend on any sensor in which the neural network functions as the observers required for motor control.

For the above, the problem of starting the BLCD motor must be considered because, as it does not have any sensor, the observers will not give results at speeds close to zero, therefore it implies a greater challenge.

Author Contributions: Investigation, writing—original draft preparation, software, and validation, O.-D.R.-C.; Conceptualization, Project administration, and writing—review and editing, F.T.-R. All authors have read and agreed to the published version of the manuscript.

Funding: This research received no external funding.

Conflicts of Interest: The authors declare no conflict of interest.

References

1. Zhao, M.; Liu, X.; Su, H. Robust adaptive speed control of disturbed brushless direct current motor. In Proceedings of the 2017 Eighth International Conference On Intelligent Control and Information Processing (ICICIP), Hangzhou, China, 3–5 November 2017; pp. 141–146. [CrossRef]
2. Shanmugasundram, R.; Zakaraiah, K.M.; Yadaiah, N. Modeling, simulation and analysis of controllers for brushless direct current motor drives. *J. Vib. Control* **2013**, *19*, 1250–1264. [CrossRef]
3. Arulmozhiyal, R.; Kandiban, R. Design of fuzzy PID controller for brushless DC motor. In Proceedings of the 2012 International Conference on Computer Communication and Informatics, Coimbatore, India, 10–12 January 2012; pp. 1–7. [CrossRef]
4. Premkumar, K.; Manikandan, B. Adaptive neuro-fuzzy inference system based speed controller for brushless DC motor. *Neurocomputing* **2014**, *138*, 260–270. [CrossRef]
5. Al-Maliki, A.Y.; Iqbal, K. FLC-based PID controller tuning for sensorless speed control of DC motor. In Proceedings of the 2018 IEEE International Conference on Industrial Technology (ICIT), Lyon, France, 20–22 February 2018; pp. 169–174. [CrossRef]
6. Mamadapur, A.; Mahadev, G.U. Speed Control of BLDC Motor Using Neural Network Controller and PID Controller. In Proceedings of the 2019 2nd International Conference on Power and Embedded Drive Control (ICPEDC), Chennai, India, 21–23 August 2019; pp. 146–151. [CrossRef]
7. Liu, L.; Liu, Y.J.; Chen, C.P. Adaptive neural network control for a DC motor system with dead-zone. *Nonlinear Dyn.* **2013**, *72*, 141–147. [CrossRef]
8. Ibrahim, H.; Hassan, F.; Shomer, A.O. Optimal PID control of a brushless DC motor using PSO and BF techniques. *Ain Shams Eng. J.* **2014**, *5*, 391–398. [CrossRef]
9. Ramya, A.; Balaji, M.; Kamaraj, V. Adaptive MF tuned fuzzy logic speed controller for BLDC motor drive using ANN and PSO technique. *J. Eng.* **2019**, *2019*, 3947–3950. [CrossRef]
10. Potnuru, D.; Mary, K.A.; Babu, C.S. Experimental implementation of Flower Pollination Algorithm for speed controller of a BLDC motor. *Ain Shams Eng. J.* **2019**. [CrossRef]
11. Wang, M.S.; Chen, S.C.; Shih, C.H. Speed control of brushless DC motor by adaptive network-based fuzzy inference. *Microsyst. Technol.* **2018**, *24*, 33–39. [CrossRef]
12. Templos-Santos, J.L.; Aguilar-Mejia, O.; Peralta-Sanchez, E.; Sosa-Cortez, R. Parameter Tuning of PI Control for Speed Regulation of a PMSM Using Bio-Inspired Algorithms. *Algorithms* **2019**, *12*, 54. [CrossRef]
13. Merugumalla, M.K.; Kumar, N.P. FFA-based speed control of BLDC motor drive. *Int. J. Intell. Eng. Inform.* **2018**, *6*, 325–342. [CrossRef]

14. ELkholy, M.M.; El-Hay, E.A. Efficient dynamic performance of brushless DC motor using soft computing approaches. *Neural Comput. Appl.* **2019**, 1–14. [CrossRef]
15. Yu, J.; Shi, P.; Dong, W.; Chen, B.; Lin, C. Neural network-based adaptive dynamic surface control for permanent magnet synchronous motors. *IEEE Trans. Neural Netw. Learn. Syst.* **2014**, *26*, 640–645. [CrossRef] [PubMed]
16. Cheng, J.; Zhang, G.; Lu, C.; Wu, C.; Xu, Y. Research of brushless DC motor control system based on RBF neural network. In Proceedings of the 2017 32nd Youth Academic Annual Conference of Chinese Association of Automation (YAC), Hefei, China, 19–21 May 2017; pp. 521–524. [CrossRef]
17. Abed, W.; Sharma, S.; Sutton, R. Diagnosis of bearing fault of brushless DC motor based on dynamic neural network and orthogonal fuzzy neighborhood discriminant analysis. In Proceedings of the 2014 UKACC International Conference on Control (CONTROL), Loughborough, UK, 9–11 July 2014; pp. 378–383. [CrossRef]
18. Luo, S.; Wu, S.; Gao, R. Chaos control of the brushless direct current motor using adaptive dynamic surface control based on neural network with the minimum weights. *Chaos Interdiscip. J. Nonlinear Sci.* **2015**, *25*, 073102. [CrossRef] [PubMed]
19. Saleh, A.L.; Obed, A.A.; Qasim, H.H.; Breesam, W.I.; Al-Yasir, Y.I.; Parchin, N.O.; Abd-Alhameed, R.A. Wavelet Neural Networks for Speed Control of BLDC Motor. In *Energy Dissipation*; IntechOpen: London, UK, 2020. [CrossRef]
20. Naung, Y.; Anatolii, S.; Lin, Y.H. Speed Control of DC Motor by Using Neural Network Parameter Tuner for PI-controller. In Proceedings of the 2019 IEEE Conference of Russian Young Researchers in Electrical and Electronic Engineering (EIConRus), Saint Petersburg/Moscow, Russia, 28–31 January 2019; pp. 2152–2156. [CrossRef]
21. Kim, C.H. Artificial Intelligent Control for DC Motor via Dynamic Neural Networks. *J. Korean Soc. Railw.* **2019**, *22*, 467–472. [CrossRef]
22. Ho, T.Y.; Chen, Y.J.; Chen, P.H.; Hu, P.C. The design of a motor drive based on neural network. In Proceedings of the 2017 International Conference on Applied System Innovation (ICASI), Sapporo, Japan, 13–17 May 2017; pp. 337–340. [CrossRef]
23. Khadar, S.; Kouzou, A.; Rezzaoui, M.M.; Hafaifa, A. Sensorless control technique of open-end winding five phase induction motor under partial stator winding short-circuit. *Period. Polytech. Electr. Eng. Comput. Sci.* **2020**, *64*, 2–19. [CrossRef]
24. König, N.; Nienhaus, M.; Grasso, E. Analysis of Current Ripples in Electromagnetic Actuators with Application to Inductance Estimation Techniques for Sensorless Monitoring. *Actuators* **2020**, *9*, 17. [CrossRef]
25. Che, H.; Wu, B.; Yang, J.; Tian, Y. Speed sensorless sliding mode control of induction motor based on genetic algorithm optimization. *Meas. Control* **2020**, *53*, 192–204. [CrossRef]
26. Wu, S.; Zhang, J. A terminal sliding mode observer based robust backstepping sensorless speed control for interior permanent magnet synchronous motor. *Int. J. Control Autom. Syst.* **2018**, *16*, 2743–2753. [CrossRef]
27. Aguilar, L.; Ramírez-Villalobos, R.; Ferreira de Loza, A.; Coria, L. Robust sensorless speed tracking controller for surface-mount permanent magnet synchronous motors subjected to uncertain load variations. *Int. J. Syst. Sci.* **2020**, *51*, 35–48. [CrossRef]
28. Kivanc, O.C.; Ozturk, S.B. Low-Cost Position Sensorless Speed Control of PMSM Drive Using Four-Switch Inverter. *Energies* **2019**, *12*, 741. [CrossRef]
29. Urbanski, K.; Janiszewski, D. Sensorless Control of the Permanent Magnet Synchronous Motor. *Sensors* **2019**, *19*, 3546. [CrossRef]
30. Wang, B.; Wang, Y.; Feng, L.; Jiang, S.; Wang, Q.; Hu, J. Permanent-Magnet Synchronous Motor Sensorless Control Using Proportional-Integral Linear Observer with Virtual Variables: A Comparative Study with a Sliding Mode Observer. *Energies* **2019**, *12*, 877. [CrossRef]
31. Shiva, B.S.; Verma, V.; Khan, Y.A. Q-MRAS-based speed sensorless permanent magnet synchronous motor drive with adaptive neural network for performance enhancement at low speeds. In *Innovations in Soft Computing and Information Technology*; Springer: Singapore, 2019; pp. 103–116. [CrossRef]
32. Elbeji, O.; Hannachi, M.; Benhamed, M.; Sbita, L. Artificial neural network-based sensorless control of wind energy conversion system driving a permanent magnet synchronous generator. *Wind Eng.* **2020**, 0309524X20903252. [CrossRef]

33. Sreeram, K. Design of fuzzy logic controller for speed control of sensorless BLDC motor drive. In Proceedings of the 2018 International Conference on Control, Power, Communication and Computing Technologies (ICCPCCT), Kannur, India, 23–24 March 2018; pp. 18–24. [CrossRef]
34. Saed, N.; Mirsalim, M. Enhanced sensor-less speed control approach based on mechanical offset for dual-stator brushless DC motor drives. *IET Electr. Power Appl.* **2020**, *14*, 885–892. [CrossRef]
35. Vanchinathan, K.; Valluvan, K. A metaheuristic optimization approach for tuning of fractional-order PID controller for speed control of sensorless BLDC motor. *J. Circuits Syst. Comput.* **2018**, *27*, 1850123. [CrossRef]
36. Verma, V.; Pal, N.S.; Kumar, B. Speed Control of the Sensorless BLDC Motor Drive Through Different Controllers. In *Harmony Search and Nature Inspired Optimization Algorithms*; Springer: Singapore, 2019; pp. 143–152. [CrossRef]
37. Rif'an, M.; Yusivar, F.; Kusumoputro, B. Sensorless-BLDC motor speed control with ensemble Kalman filter and neural network. *J. Mechatron. Electr. Power Veh. Technol.* **2019**, *10*, 1–6. [CrossRef]
38. Veras, F.C.; Lima, T.L.V.; Souza, J.S.; Ramos, J.G.G.S.; Lima Filho, A.C.; Brito, A.V. Eccentricity Failure Detection of Brushless DC Motors From Sound Signals Based on Density of Maxima. *IEEE Access* **2019**, *7*, 150318–150326. [CrossRef]
39. Medeiros, R.L.V.; Filho, A.C.L.; Ramos, J.G.G.S.; Nascimento, T.P.; Brito, A.V. A Novel Approach for Speed and Failure Detection in Brushless DC Motors Based on Chaos. *IEEE Trans. Ind. Electron.* **2019**, *66*, 8751–8759. [CrossRef]
40. Garcia-Calva, T.A.; Morinigo-Sotelo, D.; Garcia-Perez, A.; Camarena-Martinez, D.; de Jesus Romero-Troncoso, R. Demodulation Technique for Broken Rotor Bar Detection in Inverter-Fed Induction Motor Under Non-Stationary Conditions. *IEEE Trans. Energy Convers.* **2019**, *34*, 1496–1503. [CrossRef]
41. Delgado-Arredondo, P.A.; Morinigo-Sotelo, D.; Osornio-Rios, R.A.; Avina-Cervantes, J.G.; Rostro-Gonzalez, H.; de Jesus Romero-Troncoso, R. Methodology for fault detection in induction motors via sound and vibration signals. *Mech. Syst. Signal Process.* **2017**, *83*, 568–589. [CrossRef]
42. Zamudio-Ramirez, I.; Osornio-Rios, R.A.; Trejo-Hernandez, M.; Romero-Troncoso, R.d.J.; Antonino-Daviu, J.A. Smart-Sensors to Estimate Insulation Health in Induction Motors via Analysis of Stray Flux. *Energies* **2019**, *12*, 1658. [CrossRef]
43. Zhou, Z.; Li, S.; Zhou, Y.; Jiao, Y. Simulation of BLDC in Speed Control System on PSIM and Matlab/Simulink Co-simulation Platform. In Proceedings of the First, Symposium on Aviation Maintenance and Management, Xi'an, China, 25–28 November 2013; pp. 621–629. [CrossRef]
44. Mozaffari Niapour, S.; Shokri Garjan, G.; Shafiei, M.; Feyzi, M.; Danyali, S.; Bahrami Kouhshahi, M. Review of permanent-magnet brushless DC motor basic drives based on analysis and simulation study. *Int. Rev. Electr. Eng.* **2014**, *9*, 930–957. [CrossRef]
45. Frolov, V.Y.; Zhiligotov, R.I. Development of sensorless vector control system for permanent magnet synchronous motor in Matlab Simulink. *J. Min. Inst.* **2018**, *229*, doi:10.25515/PMI.2018.1.92. [CrossRef]
46. Nautiyal, C.T.; Singh, S.; Rana, U. Recognition of noisy numbers using neural network. In *Soft Computing: Theories and Applications*; Springer: Singapore, 2018; pp. 123–132. [CrossRef]
47. Ramirez-Leyva, F.; Trujillo-Romero, F.; Caballero-Morales, S.; Peralta-Sanchez, E. Direct Torque Control of a Permanent-Magnet Synchronous Motor with Neural Networks. In Proceedings of the 2014 International Conference on Electronics, Communications and Computers (CONIELECOMP), Cholula, Mexico, 26–28 February 2014; pp. 71–76. [CrossRef]
48. Trujillo-Romero, F.; Jimenez Hernandez, J.d.C.; Gomez Lopez, W. Predicting electricity consumption using neural networks. *IEEE Lat. Am. Trans.* **2011**, *9*, 1066–1072. [CrossRef]
49. Badillo, H.Y.; Olvera, R.T.; Mejía, O.A.; Carbajal, F.B. Control Neuronal en Línea para Regulación y Seguimiento de Trayectorias de Posición para un Quadrotor. *Revista Iberoamericana de Automática e Informática industrial* **2017**, *14*, 141–151. [CrossRef]
50. Wu, S.; Liu, J. Simulation Analysis of Dynamic Characteristics of AC Motor Based on BP Neural Network Algorithm. In Proceedings of the International Conference on Cyber Security Intelligence and Analytics, Shenyang, China, 21–22 February 2019; pp. 277–286. [CrossRef]

51. Sadrossadat, S.A.; Rahmani, O. ANN-based method for parametric modelling and optimising efficiency, output power and material cost of BLDC motor. *IET Electr. Power Appl.* **2020**, *14*, 951–960. [CrossRef]
52. Mishra, P.; Banerjee, A.; Ghosh, M. FPGA Based Real-Time Implementation of Quadral-Duty Digital PWM Controlled Permanent Magnet BLDC Drive. *IEEE ASME Trans. Mechatron.* **2020**. [CrossRef]

© 2020 by the authors. Licensee MDPI, Basel, Switzerland. This article is an open access article distributed under the terms and conditions of the Creative Commons Attribution (CC BY) license (http://creativecommons.org/licenses/by/4.0/).

MDPI
St. Alban-Anlage 66
4052 Basel
Switzerland
Tel. +41 61 683 77 34
Fax +41 61 302 89 18
www.mdpi.com

Mathematical and Computational Applications Editorial Office
E-mail: mca@mdpi.com
www.mdpi.com/journal/mca

www.ingramcontent.com/pod-product-compliance
Lightning Source LLC
LaVergne TN
LVHW070906170726
843515LV00005B/715

* 9 7 8 3 0 3 6 5 1 4 5 2 9 *